Ecology and Environment : Current Perspectives

Ecology and Environment : Current Perspectives

By

H.K. Pathak

CYBER TECH PUBLICATIONS
4264/3, Ansari Road, Daryaganj, New Delhi-110002 (India)
Ph.: 011-23244078. Fax: 011-23280028
E-Mail: cyberpublicationsdelhi@yahoo.com

ECOLOGY AND ENVIRONMENT : CURRENT PERSPECTIVES

H.K. Pathak

First Edition 2009

Published by :

G.S. Rawat for Cyber Tech Publications
4264/3, Ansari Road, Daryaganj, New Delhi-110002 (India)
Ph.: 011-23244078. Fax: 011-23280028
E-Mail: cyberpublicationsdelhi@yahoo.com

Printed at : **Amit Enterprises**, Maujpur, Delhi, (9810878184)

Preface

The decade of the 1970s was marked by a growing climate consciousness, both popular and scientific. The new interest was sparked by a series of extreme climate events and related disruptions, and by scientific speculation as to increased climate variability and possible climate change. Two sets of events during this period attracted both scientific and public interest. The first, in 1972, was the apparent simultaneous occurrence of unfavorable weather in many parts of the globe and its speculative relationship to a wide variety of socioeconomic events, including the quadrupling of various commodity prices around the world, food shortages in the Sahel of West Africa and in South Asia, a drastic fall in the anchovy fishery of the Pacific, and even changes in government in Ethiopia and Niger. The second was an emerging scientific consensus that human-induced alterations in the chemical constituents of the atmosphere could lead to large regional, and even global, changes of the atmosphere in the form of more acidic rain, greater ultraviolet radiation, and altered temperatures.

In sum, the diversity of novel climatic experience continues unabated, the recognition of potential sources of human-induced alteration has increased, and the pace and degree of change are questioned and debated. Nonetheless, it is widely agreed that one such change, a long-term global warming derived from the enrichment of the atmospheric content of the `greenhouse' gases, is underway. Within the time period of the projected global average warming, measured in tens of hundreds of years, sustained variation of climate will occur in many places, and lesser periods of favorable or unfavorable climate will occur in most places a function of normal variability. Where these changes are large the extremes greater than what is customary-where people and places are vulnerable, or where human activity meshes poorly with natural opportunity, significant climate impacts to people, ecosystems and societies are likely to occur. How to respond to such impacts adjusting to changing climate, coping with extremes, matching human needs to climate endowment are issues of considerable import. The scientific study of climate and society will inform societal response. Concepts of climate impact assessment are new, the methods are still under development. This book is an authoritative review of these methods and concepts, a contribution of the Scientific Committee on Problems of the Environment to the World Climate Program.

—Author

Contents

Contents

1

Ecology and Environment: Introduction

INTRODUCTION

Environmental conservation is an integral part of the socio economic development. The growing population, high degree of mechanisation and steep rise in energy use has led to activities that directly or indirectly affect the sustainability of the environment. The various kinds of pollution/pollutants may be broadly categorized as (a) Air pollution, (b) Water pollution, (c) Solid waste and (d) Industrial and hazardous waste. The sustainable use of bio-diversity is fundamental to ecologically sustainable development. India is one of the 12 mega diversity countries of the world.

The survival and well being of any nation depends on sustainable social and economic progress that satisfies the needs and aspiration of the present without compromising the interest of future generations. Environmental problems are largely the by-products of affluence marked by resource wasteful life style. Recycling of solid and liquid wastes, bio-composting, tree planting, etc. are important for environmental conservation.

ECOLOGICAL SECURITY

Spiraling population and increasing industrialisation are posing a serious challenge to the preservation of the terrestrial and aquatic ecosystems. Environmental protection is the key to ensuring a healthy life for the people. Environmental problems are on the increase and are more pronounced in densely populated cities. Creation of awareness regarding the ecological hazards among the public is absolutely essential. Environmental conservation and abatement of pollution are critical for sustainable development.

Department of Environment

The Department of Environment was created in 1995 as the nodal Department for dealing with environmental management of the State.

The Department is entrusted with the implementation of major projects like pollution abatement in the Cauvery, Vaigai and Tamiraparani rivers, pollution abatement in Chennai city waterways, National Lake Conservation Programme and all aspects of environment other than those dealt with by Tamil Nadu Pollution Control Board.

One of the main objectives of the department is to implement Environmental Awareness Programme Wide publicity is being given on World Environment Day, Ozone day and on Bhogi Day to create environmental awareness among the general public. Due to the concerted efforts, the level of air and noise pollution has been brought down to the tune of 20-25% in the last three years.

To create environmental awareness among the school and college studies, 1260 eco-clubs have been formed in all the districts of State involving selected educational institutions and NGOs. It is proposed to strengthen these existing eco-clubs. The outstanding NGOs, experts and individuals are honoured with environmental awards in recognition of their excellent contribution in the field of environment.

Tamil Nadu Pollution Control Board

The Tamil Nadu Pollution Control Board enforces the provisions of the Water (Prevention and Control of Pollution) Act, 1974 as amended, the Water (Prevention and Control of Pollution) Cess Act, 1977 as amended, the Air (Prevention and

Control of Pollution) Act, 1981 as amended and the relevant provisions/rules of the Environment (Protection) Act, 1986 to prevent, control and abate pollution and for protection of environment.

The Board functions with its Head Office at Chennai. There are 25 District Offices and 14 Environmental Laboratories established by the Board.

Monitoring of Industries

The Board has inventorised about 28,000 industries. The Board has prescribed standards for discharge of effluent, ambient air quality and gaseous emissions from various industries and the industries have to take necessary pollution control measures to meet the standards prescribed by the Board. For effective monitoring, the Board has classified the industries into red, orange and green, based on their pollution potential.

Procedure for Issue of Consent

The Board issues consent to industries in two stages under the Water Act and the Air Act for establishment and operation of industrial units. Consent to establish is issued depending upon the suitability of the site, before the industry takes up the construction activity. Consent to operate is issued after installation of effluent treatment plant and air pollution control measures, before commissioning production. Consent is issued subject to general conditions and specific conditions.

Common Effluent Treatment Plants

The Board plays an important role in the establishment of Common Effluent Treatment Plants (CETPs) for clusters of small polluting industries in various parts of the State. Tamil Nadu is a pioneering State in India in establishing CETPs. So far, proposals for 50 CETPs have been formulated. Of these, 33 CETPs are under operation and 17 CETPs are under various stages of implementation. Towards the implementation of CETPs, State subsidy is granted by Government of Tamil Nadu, upto 25% of the project cost and Central subsidy is granted by

Government of India, upto 25% of the project cost. The details of CETPs are as follows:

Sl. No.	Sector	No. of CETPs formed	No. of CETPs under operation
1	Tanneries	24 schemes	14 schemes
2	Textile Bleaching and Dyeing	25 schemes	18 schemes
3	Hotels and Lodging Houses	1 scheme	1 scheme
	Total	50 schemes	33 schemes

Cleaner Technologies & Water and Energy Conservation

With active support and encouragement from the Board, industrial units in Tamil Nadu have started switching over to cleaner technologies and also started water and energy conservation measures.

Air quality monitoring As per the provisions of the Air Act, the entire State of Tamil Nadu has been declared as air pollution control area. The Board is monitoring the ambient air quality in Chennai (3 stations), Coimbatore (3 stations), Thoothukudi (3 stations), Madurai (3 stations) and Salem (1 station) under the National Ambient Air Quality Monitoring Programme. Under the State Ambient Air Quality Monitoring Programme, the Board has established 5 ambient air quality monitoring stations in Chennai city and 5 in Tiruchirapalli. Towards preparation of the environmental management plan for Chennai city, ambient air quality surveys have been conducted at 41 stations in Chennai to identify the most sensitive locations with respect to air pollution. The Board has established 6 continuous ambient air quality monitoring systems at Cuddalore, Thoothukudi, Ranipet, Manali-Thiruvallur, Royapuram-Chennai, Kathivakkam-Thiruvallur to assess the level of pollutants in the ambient air and the adequacy of air pollution control measures provided by the industries in the vicinity.

Vehicle Emission Monitoring

The Board is carrying out the vehicle emission monitoring in Chennai, Dindigul, Palani, Udhagamandalam and

Chengalpattu. In addition, private agencies have been authorised by the Transport Department in Chennai city to check the emission level of the vehicles. The Board has upgraded and computerised all its vehicle emission monitoring stations for testing diesel driven vehicles. The Transport Corporations have also been instructed to closely monitor the emission levels of their buses.

For controlling vehicular emission, cleaner fuel like unleaded petrol, petrol with 3% benzene and low sulphur fuel (0.05%) have been introduced in Chennai Metropolitan Area. Passenger cars complying with Bharat stage-II norms alone are registered in Chennai since July 2001. 2T oil auto dispensing system have been provided in retail outlets.

The Board is also participating in a research project with an non governmental organisation and the Civil Supplies Department to study the use of gas chromatograph to detect fuel adulteration. Action has already been taken to introduce auto liquefied petroleum gas in Chennai as it is a cleaner fuel. Steps are being taken to popularise the use of liquefied petroleum gas for autorickshaws, call taxis and other private vehicles which will help in improving air quality.

(1) Vehicle Emission Monitoring Stations provided by the Board. 8 Nos.

(2) Vehicle Emission Monitoring Stations provided by the Board in MTC Depots. 6 Nos.

(3) Emission Checking Stations provided by Private Agencies. 236 Nos.

(4) Auto Liquefied Petroleum Gas Dispensing Stations commissioned in Chennai. 12 Nos.

Noise Level Monitoring

Towards controlling noise pollution in urban areas, about 52,586 air horns were removed as of December 2004 from buses and lorries throughout the State. All the districts have been declared as air horn free districts. For noise level monitoring at the district level, sophisticated noise level meters have been provided to the District Offices of the Board.

Water Quality Monitoring

Pollution of major rivers in the State is caused by the discharge of untreated sewage from the urban local bodies and panchayats and untreated or partially treated effluent from industries. In case of industrial pollution, it is the responsibility of the industrial units to provide the required effluent treatment plants either individually or collectively so as to achieve the standards. Various pollution abatement schemes are being implemented under the National River Conservation Programme under the coordination of the Department of Environment.

Water Quality Monitoring Programmes

Under the Global Environmental Monitoring System, the Board is closely monitoring the quality of water in the Cauvery basin at Mettur, Pallipalayam, Musiri and ground water quality at Musiri. Similarly, water quality of rivers Cauvery (16 stations), Tamiraparani (7 stations), Palar (1 station) and Vaigai (1 station) and the three important lakes in Udhagamandalam, Kodaikkanal and Yercaud are being monitored under the Monitoring of Indian National Aquatic Resources System by the Board. The Board is continuously monitoring the Chennai city water ways to prevent pollution due to discharge of trade effluent from industries and sewage from local bodies and is collecting and analysing samples of river water and outfalls at regular intervals, since 1991.

WASTES MANAGEMENT

Hazardous Wastes Management

The Board has identified and inventorised 2,177 units generating hazardous wastes and compiled a computerised data base of hazardous waste streams in respect of existing units as per the Hazardous Wastes (Management and Handling) Rules, 1989 as amended in 2000 and 2003.

Municipal Solid Waste Management

With increasing urbanisation and rising levels of municipal solid wastes, an urgent task is to evolve scientific approaches to segregate, handle and dispose the solid wastes in urban areas. Tiruppur municipality has started segregation of waste at source and has engaged a private agency for composting the segregated waste. All the municipalities have started segregation and collection of waste at source and the composting work has begun. The Board has issued no objection certificates to 56 municipalities and to one corporation for their site identified for composting of municipal solid wastes. Out of this, authorizations have been issued to 11 municipalities to commence their composting works.

Plastic Wastes Management

Realizing the environmental problems caused by indiscriminate use and disposal of plastics, awareness has been created regarding reduction in the use of throwaway plastics through various campaigns. Exhibitions on eco-friendly alternatives to plastics have been arranged in Chennai and Districts. Billboards educating people against the use and disposal of throwaway plastics were placed on Metropolitan Transport Corporation buses, in Chennai. Awareness campaigns are conducted in tourist towns and other pilgrim centres. As an alternative to plastics, 12 training programmes for the production of palm leaf products have been conducted to self help groups through the Central Palmgur and Palm Products Institute of Village Industries Commission.

Biomedical Wastes Management

The Board enforces the Biomedical Waste (Management and Handling) Rules, 1998 as amended in 2000. Towards establishing common biomedical waste treatment and disposal facilities for the private health care units, 11 sites have been identified by the Indian Medical Association. These 11 common facilities have been issued authorization out of which 6 common facilities in Kancheepuram district (2 facilities), Vellore district, Thanjavur district, Ramanathapuram district and the Nilgiris district are under operation. 5 common facilities in Coimbatore district (2 facilities), Salem district, Virudhunagar distirct and Tirunelveli district are under establishment.

OTHER ACTIVITIES OF THE BOARD

Applied Research and Development

The Board has established an applied research and development wing in the Advanced Environmental Laboratory at Chennai for developing cleaner technologies that would reduce the generation of pollution. The Board has established a research fund from cess fund of the Board to assist various institutions for conducting research studies in the field of environmental protection, pollution control and development of cleaner technologies.

Environmental Training Institute

The Board has established an Environmental Training Institute with Danish assistance of Rs.4 crores at its head office to impart training to industries, local bodies, non governmental organisations and the staff of the Board to enable to monitor and advise on pollution abatement and prevention techniques.

Year	No. of Training Programmes Conducted	No. of Participants
2002-2003	18	2,548
2003-2004	24	1,674

Environmental Awareness

An environmental pavilion has been set up at Periyar Science and Technology Centre, Chennai wherein various painting, essay writing and oratorical contests on environmental issues are conducted and video films and short films are screened. The Board has established a pollution awareness and assistance cell in the Corporate office. Environmental awareness programmes are being conducted regularly. Special awareness campaigns are conducted against air and noise pollution during festival seasons such as Deepavali and Bhogi. These campaigns have been quite successful. During the year 2004-2005, the Board with the District Collectors has taken up the activity of promoting eco-friendly pilgrim centers and for creating awareness regarding various environmental issues engaging non governmental organisations, environmental activists and volunteers.

Green Awards

Green awards are given to the District Collectors based on the assessment of the personal contribution of the Collectors to the promotion of environmental protection and sustainable development. Green awards were also given to the municipal commissioners. Green Innovators Award was given during the year 2003 to an industrial unit, which has contributed innovatively in tackling a major industrial pollution problem in the textile processing industry.

Complaint Cell

An active 'Complaint cell' is functioning in the corporate office, Chennai to take appropriate action against complaints received from various individuals/organisation/NGOs etc., regarding water, air, noise and vehicular pollution.

Preparation of Environmental Atlas

The Board with assistance of the Central Pollution Control Board is implementing the spatial environmental planning programme for preparing district environmental atlas in respect of Thiruvallur, Kancheepuram, Coimbatore, Vellore and Thoothukudi districts. The environmental management plan for Chennai city has also been prepared and submitted for finalisation to the Central Pollution Control Board.

Tenth Five Year Plan 1. Objective

The goals / objectives for the State's Tenth Five Year Plan period were specified as follows:-

- o Protection of environment and reduction of wastes at source, besides continuing enforcement and monitoring activities
- o Conservation and enhancement of life support systems like land, water, forests, biodiversity, ocean and the atmosphere giving ecological security

- o Properly treating or converting the pollutants into useful products and thereby reducing their negative impact
- o Promotion of suitable measures and technologies for recycling/reusing
- o Emphasising environmentally safe waste disposal options viz., composing of all organic wastes, common facilities for bio-medical wastes etc.
- o Maintaining essential ecological processes to ensure genetic diversity and prevent environmental degradation.
- o Promoting the active participation of people at all levels-from the Village Panchayat to the State as a whole-in the preservation of the environment and promotion of sustainable development through awareness campaigns.

Outlay

The outlay for "Ecology and Environment" sector for the Tenth Plan Period was Rs.197.20 crores, out of which the State's share outlay is Rs.113.05 crores. The details are given below.

The sources of funding for the Tamil Nadu Pollution Control Board and the Department of Environment are also indicated. (Rs. in crores)

Sl. No.	Dept. / Agency	State fund	Central fund	Fund from Internal/own mobilisation (Board) resources	Total
1.	Tamil Nadu Pollution Control Board	42.02	14.40	---	56.42
2.	Department of Environment	113.05	27.73	----	140.78
	Total	113.05	69.75	14.40	197.20

Annual Plan 2002-03

The actual expenditure incurred during the year 2002-03 for the Ecology and Environment was Rs.51.22 lakhs as against the budget provision of Rs.37.43 lakhs.

Annual Plan 2003-04

An amount of Rs.97.41 lakhs was budgeted for the year 2003-04. Against this, the actual expenditure incurred was Rs.75.69 lakhs. E: Annual Plan 2004-05

An amount of Rs.82.92 lakhs was budgeted for the year 2004-05. Against this, the expenditure anticipated is Rs.92.18 lakhs. The performance of the new schemes sanctioned in 2004-05 is as follows:

Environmental Education-Clean and Green Streets

A pilot project has been initiated comprising of 40 school eco clubs in Chennai to adopt one street adjacent to their school premises under this scheme. The solid waste generated in the adopted streets will follow source segregation. Biodegradable wastes will be composted by forming aerobic composting/ vermi composting pits in the school premises. The natural manure prepared will be used to fertilise tree saplings planted along the adopted streets.

Thus the streets adopted by Eco-club will become clean and green. The eco-club members will disseminate the message of environmental conservation. The teacher coordinator of the eco-club is incharge of the programme at the school level. An assistance of Rs.18,750/-per school is given under the programme and the entire amount of Rs.7.50 lakhs has been spent during 2004-2005.

Environment Research and Development Programme An amount of Rs.10.00 lakhs was provided for environment research and development programme. Innovative Environmental Research projects have been promoted in the areas of bio-remediation of lakes and remediation of effluent affected sites using appropriate tree species. A Rapid Environmental Impact Assessment of Tsunami affected coastal and marine areas has been undertaken during February-March 2005 and a sum of Rs.5.13 lakhs has been spent during 2004-2005.

Establishment of Water Quality Monitoring Laboratory

The Department of Environment has undertaken cleaning of the rivers and lakes in Tamil Nadu under the National River Conservation Plan (NRCP) and the National Lake Conservation Plan (NLCP). It is important to undertake a baseline survey and also monitor the water quality in the subsequent years. It is proposed to establish a small water quality monitoring lab in the Department of Environment for exclusively monitoring the water quality of these rivers and lakes by testing the samples effectively. A sum of Rs.10.00 lakhs has been provided for the above programme and this will be utilized during 2005-2006.

Conservation of Coastal Ecosystem

Coastal area forms a highly complex and dynamic ecosystem, sensitive to developmental pressures such as industrialization. These could be minimized through prudent Coastal Zone Management practices.

An Integrated Coastal Zone

Management Plan involving various departments, agencies and all the stakeholders will be prepared for the Pulicat to Palar stretch of Tamil Nadu coast during the Tenth Plan. A sum of Rs.10.00 lakhs has been provided for this proposal.

The Department of Environment is also implementing CRZ notification 1991 under the Environment Protection Act 1986 and conservation of coast in line with this notification is quite essential. Government of India have emphasized that violations along the coast should be checked by the State Government through increased surveillance.

It is proposed to form a Green Squad for Coastal Conservation for checking CRZ violations. Government of India has sanctioned Rs.5.00 lakhs for this purpose with which a vehicle has been purchased. Out of the Government of Tamil

Nadu allocation, a sum of Rs.5.00 lakhs has been spent under this scheme during 2004-2005 and the balance of Rs.7.00 lakhs will be spent during 2005-06.

Annual Plan 2005-06

For the year 2005-06, an amount of Rs.109.96 lakhs is proposed. The details of the schemes are as follows:-

Combating Desertification of Land

Tamil Nadu has about 3.7 million ha. of land under various stages of degradation. The main areas in the districts of Ramanathapuram, Erode, Theni, Sivagangai and Nagapattinam face the problem of desertification of land. Initially, a pilot project is proposed to be taken up in Ramanathapuram district, which will include

Oasis Plantation for stabilization of sand dunes, construction of check dams and percolation ponds for soil and moisture conservation and formation of Self Help Groups for alternate employment generation. The scheme will also involve planting of legumes and grasses to support animal husbandry in these areas. This project may become a model for tackling larger areas in future under the "United Nation Convention for Combating Desertification Programme". A sum of Rs.15.00 lakhs has been proposed to carry out this project over 100 ha. during the year 2005-06. Eco-City Plan For Tiruvannamalai

The objective of the programme is to incorporate environment considerations into Urban Planning and prepare an Environmental Management Plan for improving the quality of environment by mapping the environmental profile of study area and to identify the environmental pollution hotspots. To prepare an environmental management plan that includes rehabilitation and mitigation measures and to recommend guidelines for environmentally compatible land use planning. Tiruvannamalai, being a temple city attracting lot of tourists, special emphasis must be given to the carrying capacity of the city. A sum of Rs.10.00 lakhs has been proposed to carry out this project during the year 2005-06.

Greening of Sholingur Temple Hillock

Eco restoration of the degraded temple hillock was sanctioned for 2003-2004 at a cost of Rs.9.00 lakhs. Saplings have been planted at Sholingur Hillock, which are being nurtured and maintained. Afforestation over another 50 Ha. is proposed to be taken up during this year. Moreover the temple tank at the foothills will be desilted and a Nandavanam created around it. This project has been approved to be taken up at a cost of Rs.6.00 lakhs during 2005-06.

Aesthetic Amelioration of Temple Hillock at Palani

Reversal of loss of ecology has been included as one of the components under Tenth Five Year Plan. A pilot project for aesthetic amelioration of temple hillock has already been undertaken in Sholingur Temple in Vellore District last year.

For the year 2005-06 it is proposed to take up this project in the Temple Hillock of Arulmigh Dhandayutha Pani Swamy of Palani in Dindigul District. A sum of Rs.10.00 lakhs has been budgeted for this project to be carried out during the year 2005-06.

Eco-Restoration Plan for Selected Lakes In Chennai City

There are about forty-six lakes in Chennai city. Most of these lakes are polluted with sewage and garbage and are environmentally degraded. Being in and around the city these lakes are also prone for encroachment. Chennai being a water scarce city, the eco-restoration of these lakes will go a long way in augmenting the water requirements of the city and also will be recharging the dwindling ground water table. During the monsoon season the residential areas around these lake areas get inundated due to blockages in the natural drains to the lakes.

In order to alleviate these problems, a comprehensive eco-restoration plan for selected city lakes has been approved to be formulated at a cost of Rs.10.00 lakhs during the year 2005-06.

Plan Outlay for the Year 2005-06

An amount of Rs.109.96 lakhs is proposed for "Ecology and Environment" under State Plan schemes for the year 2005-06. The details are given below:

S.N.	Scheme	Outlay (Rs. in lakhs)		Total
1.	Department of Environment- Ongoing schemes New Schemes	58.94	51.00	109.94
2.	Pollution Control Test Wing			0.01
3.	Environmental Education, Research and Awareness Programme			0.01
	Total			109.96

Centrally Sponsored Schemes

National River Conservation Programme

The centrally sponsored scheme of "Abatement of Pollution in five polluted stretches of river Cauvery in Tamil Nadu", Tiruchi, Erode, Bhawani, Kumarapalyam and Pallipalayam at an overall project cost of Rs.3627.75 lakhs is being implemented by the Department of Environment, since 1996-97. The project period is 10 years. A sum of Rs.1946.40 lakhs has been spent as on 31.01.2005. During the year 2004-05, an amount of Rs.165.92 lakhs was utilized for the above scheme and the work will be continued during 2005-06 also.

The above project is being continued as a 100% centrally sponsored scheme. Under Non core schemes, about 95% of works in respect of construction of Low Cost Sanitation facility (LCS) an River Front Development (RFD) have been completed. Under Core-schemes, Interception and Diversion as well as Sewage Treatment works are being implemented through the TWAD Board in Erode, Bhavani, Komarapalayam, Pallipalayam and Tiruchy towns. All the Works have been completed in Tiruchy, Kumarapallayam and Erode while the works in Bhavani are under progress. Approval of detailed project report is awaited in respect Pallipalayam. Chennai City River Conservation Project (CCRCP) (Outlay Re.0.01 lakh)

A project for Rs.1700.00 crores has been approved by the Government of India to clean the Chennai city waterways under the National River Conservation Programme, of which Rs.491.52 crores will be grant for intercepting and treating sewage in six Chennai City Waterways (Cooum, Buckingham Canal, Adyar, Otteri Nullah, Captain Cotton Canal and Mambalam drain) which is being implemented through Chennai Metropolitan Water Supply and Sewerage Board. The project is being implemented since 2001. About 97% of the interception and diversion work and about 95% of the STP work has been completed at an cost of Rs.29766.71 lakhs. During the year 2004-05, Rs.10411.87 lakhs was utilized and the work will be continued during 2005-06 also. National River Conservation Plan-Seven New Towns

Government of India have approved an integrated proposal for cleaning the River Cauvery, Vaigai and Tamiraparani along seven towns viz., Tiruchy-Srirangam, Thanjavur, Kumbakonam, Karur, Myladuthurai, Madurai and Tirunelveli at a total cost of Rs.575.30 crores. Of this, 50% of the cost is obtained as grant from Government of India and the rest will be met by the Government of Tamil Nadu and the local bodies and also through public contribution.

A sum of Rs.224.39 crores has been spent so far under the project upto 28.02.2005. About 56% of work in Madurai, 46% in Kumbakonam, 70% in Thanjavur, 75% in Tirunelveli, 78% in Mayiladuthurai, 84% in Karur and 53% in Tiruchy-Srirangam has been completed. During the year 2004-2005, Rs.110.46 crores were utilized for the above scheme and work will be continued during 2005-2006 also.

National Lake Conservation Programme

Environmental upgradation through bioremediation of Ooty and Kodaikanal Lakes is being undertaken under National Lake Conservation Programme (NLCP) at a total cost of Rs.688 lakhs. Government of India has released Rs.171 lakhs for Ooty Lake and Rs.200 lakhs for Kodaikanal Lake. Of this, Rs.142.50 lakhs have been utilised for Ooty Lake. Works to the tune of Rs.33.40 lakhs has been completed in Kodaikanal Lake. Due to legal tangle the work has been stopped in Kodaikanal Lake. Necessary action is being taken to vacate the stay. During the year 2004-05, Rs.0.87 lakhs and Rs.1.60 lakhs was utilised for Ooty Lake and Kodaikanal Lake respectively. The scheme will be continued during 2005-06 also.

Palar River Conservation at Ambur

A noval project, the first of its kind in the state has been prepared for Ambur Municipality for diluting the treated industrial effluent containing high Total Dissolved Solids (TDS) with treated domestic wastewater at an estimated cost of Rs.25.00 crores and sent to Government of India for approval. Now a similar project has been prepared for Ranipet and Wallajah Municipalities at a cost of Rs.25.00 crores and sent to Government of India for approval. This project will be beneficial to the people of Ambur, Ranipet and Wallajah by providing underground sewerage system and minimise pollution of the river Palar besides bringing down the TDS of the treated industrial effluent. The project will be undertaken with the 70% financial assistance of the Ministry of Environment and Forests and the rest from the contribution of the Municipality (25%) and industries(5%).

DIRECTION & ADMINISTRATION

The Department of Environment was established in the year 1990 to look after the work of Environmental Planning Research and Ecological Regeneration, with the following posts sanctioned initially under the Centrally Assisted Scheme, namely; 'Financial Assistance to the State Department of Environment for strengthening Technical set-ups' which have now become part of State Plan:-

1. Scientist 'SE' 1
2. Scientist 'SC' 1
3. Senior Assistant 1
4. Sr.Scale Stenographer 1
5. Junior Assistant 1

6. Clerk 1
7. Driver 1
8. Attendant 2

The Department of Environment aims to protect the environment in order to promote sustainable development and prevention and control of pollution through:-

1. Eco-regeneration.
2. Environment Education, Awareness Training and Information.
3. Promotion of environmental research and training.
4. Policies & Legislations affecting environment.

The Department is responsible for the coordination of the activities of various departments/agencies and to plan a comprehensive Integrated programme for environment protection and its improvement and also responsible for exercising the administrative control on the Chandigarh Pollution Control Committee, Chandigarh for proper implementation of provisions of various Environment & Pollution Control Acts/Rules. The Department organizes seminar, training, workshops on environmental related issues from time to times besides publishing of documents/information/ pamphlets on environ-ment for public use.

ENVIRONMENTAL EDUCATION, TRAINING & INFORMATION

This is an on going scheme to impart education, training and information about environment.

This will include development of library of the department through addition of books/video films/audio-visual slides, CD/ tapes, subscription of Journals/magazines/newspapers/bulletins, membership of societies/institutions/organizations working on environment and related issues, availing/subscribing internet/ NIT terminal facilities, purchase of computers and accessories so as to be in touch with latest developments and to equip the library with facilities for keeping records/data.

It will include organizing seminars/workshops/film & audio-visual shows/exhibitions/training programmes/essay, painting, quiz and debate competitions on environment related issues.

It will also include development of public information centre with facilities as TV/Cameras, V.C.R./ Projectors/Computer/ Multimedia and related accessories to provide first hand information, Printing & publishing of documents/information brochures/pamphlets/posters for distribution.

INSTITUTIONAL SUPPORT & PUBLIC PARTICIPATION:

The environmental awareness activities/programme shall be undertaken in coordination with the Eco-Clubs/Environment Societies/NGOs whereas department shall provide only technical/financial assistance and resource material while the planning and organization of the activities/programmes/events will be left to the Eco-clubs/Environment Societies itself. The Eco-clubs have been formed in various Govt./Govt.Aided/ Recognized/Private Schools and in Navodaya/Kendriya Vidyalayas of U.T. Chandigarh whereas Environment Societies have been formed in various Colleges/Educational institutions. In addition NGOs/Societies registered under Societies Act are also considered under the scheme.

The Department will also distribute and provide facilities in way of equipments/publications/books/ teaching-learning aids/ films and support financially to organize camps/tours/excursion for creating environmental awareness amongst the students/ teachers The Department will conduct regular tours of the School children (Eco-clubs) to the wilderness areas like Botanical Garden, Sukhna Wildlife Sanctuary etc., thrice a week. The Department shall collect and disseminate information among the Eco-Clubs/Environment Societies by developing information brochures and books.

RESEARCH AND DEVELOPMENT:

The Department serves as nodal agency for environmental planning and coordination in the U.T. of Chandigarh. In order to formulate policies and plans, it is essential to have upto date database. With the growing public awareness and judicial interventions as evinced through public interest litigations, increasing responsibilities have been assigned to the Environment Deptt. to undertake research/data generation activities on the burning environmental issues and local and regional environmental problems.

The scheme will include identification/monitoring/data generation etc. on problems of air, noise pollution, water pollution, vehicular pollution, development/procurement of clean technologies & pilot plants/working models, soil and hazardous waste management, impact assessment of urbanization and industrialization and assessment of flora etc. The scheme shall be implemented through independent and or collaborative studies with reputed institutions or shall be got carried out through reputed institutions.

This will include purchase of laboratory equipments/ computers and software, expenditure for engagement of consultants/engagement of staff on contract to implement/ coordinate projects, visits to conference and seminars, undertaking trainings and financial assistance to institutions and token provision for carrying out 'Carrying Capacity Study on Chandigarh' and provisions for identification and undertaking environment impact assessment of the hazardous waste site as per the provisions of Hazardous Waste (Management & Handling) Rules, 1989.

PROTECTION & CONSERVATION OF RESOURCES

Chandigarh is situated in the shadow of the ecologically sensitive and geologically unstable shivaliks which form part of fragile Himalayan eco-systems. It is a well planned city of avenues, boulevards, gardens whose urban boundaries are defined by two seasonal rivulets. A Lake was constructed in 1958 which in due course has developed into complex eco-system. The Sukhna Lake as it is called and its adjoining areas have been declared a Wetland and the Lake itself is covered under National-Lake Conservation Plan. A detailed Project Report on Conservation of Sukhna Lake by desiltation and catchment area treatment/development with an estimate cost of Rs.73.51 crores has been prepared as per guidelines of the National River Conservation Directorate, Ministry of Environment & Forests, New Delhi and submitted to Ministry for funding.

The cost of the Project will be shared between the Ministry of Environment & Forests, New Delhi and Chandigarh Administration in the ratio of 70:30. The Project report includes desiltation of Lake, construction of soil conservation structures, vegetative methods of conserving the soil and afforestation activities in the catchment area as well as creating awareness about the ecology, flora, fauna and upgradation of lake through various programmes.

In addition the scheme will include dry desiltation of Sukhna Lake as well as wet dredging and treatment of catchment viz. soil conservation measures in Sukhna catchment, development of adjoining areas, studies to generate data on physio-chemical and biological parameters, engagement of consultants, ecological regeneration, solid waste management and public awareness and other environmental regeneration projects/activities.

ASSISTANCE TO CHANDIGARH POLLUTION CONTROL COMMITTEE

This is an ongoing scheme. The powers and the functions under the provisions of the Water (Prevention & Control of Pollution) Act, 1974 and the Air (Prevention & Control of Pollution) Act, 1981 has been delegated to Chandigarh Pollution Control Committee by the Central Pollution Control Board. The Chandigarh Pollution Control Committee is to perform various functions for the prevention, control or abatement of pollution. As per provisions of Section 35 of the Water (Prevention & Control of Pollution) Act, 1974, the State Government/U.T. Administration may provide funds to the State Pollution Control Board/Committee as the case may be in each financial year as it may think necessary to enable that Board/Committee to perform its functions under the Act. In the previous years, the Committee has improved its financial position from its own funds.

CONSTRUCTION OF PARYAVARAN BHAWAN

Chandigarh Administration decided to construct the Paryavaran Bhawan in Sector 19-B where Forest Department is presently functioning. The foundation stone of the said building was laid by H.E.Gen. (Retd.) S.F.Rodrigues, Administrator, U.T Chandigarh on the eve of World Environment Day (5th June, 2005). The proposed Paryavaran Bhawan will house the Departments of Forest, Environment, Chandigarh Pollution Control Committee & Science & Technology.

The Tata Energy Resource Institutes (TERI) has been engaged as a consultant for providing expert services for "Green Design, Energy Efficiency and Renewable Energy Systems in the Upcoming Paryavaran Bhawan". The approximate cost of the project is Rs.650 lacs.

The work of construction of Paryavaran Bhawan is likely to be completed upto the year 2008. The building will be constructed by the Engineering Department, Chandigarh Administration. The expenditure will be met under capital head "5425 Capital Outlay on Other Scientific and Environmental Research".

Apart from above, the Department of Environment is the Nodal Agency for implementing of number of other projects initiated by Govt. of India, Ministry of Environment & Forests, New Delhi from time to time.

The key motive of the schemes implemented by the Department is to create environment education and awareness with the help of Eco-Clubs/Environment Societies, environmental NGOs and other concerned as well as and protection and conservation of natural resources viz. rain water harvesting, vermin-composting, afforestation activities etc. The detail of these projects/schemes is given as under:-

National Green Corps (NGC)

The Department has established 111 Eco-Clubs in various Govt./Govt. Recognized/Private Schools Kendriya/Navodaya

Vidyalayas of U.T. Chandigarh under the National Green Corps (NGC) Programme of the Govt. of India, Ministry of Environment & Forests, New-Delhi..

The Eco-Clubs are carrying out varieties of activities/ awareness programmes viz/ rallies, marches, human chains, seminars, debates, lectures, popular talks on environmental issues, tree plantation, campus cleaning, sanitation, rainwater harvesting, solid waste management including vermin composting, etc. to spread the message of Eco-friendly/awakened Society. About 5,000 students of the Chandigarh are participating in this movement.

Strategy and Action Plan for the Conservation of Bio-Diversity

The Ministry of Environment & Forests has declared Department of Environment as Nodal Agency to formulate State Strategies and Actions Plans for conserving Bio-diversity, sustainable using biological resources in the Union Territory of Chandigarh.

Draft Report has been sent to the Govt. of India for approval. After the approval the strategies and Action Plan proposed for conserving Bio-diversity will be implemented in Chandigarh.

Preparation of Status of Environment Report of U.T.Chandigarh

The Department of Environment has been declared as Nodal Agency to coordinate with the Govt. of India, Ministry of Environment & Forests, and New Delhi with the association of Tata Energy Research Institute, New Delhi and Punjab State Council for Science & Technology "state of Environment Report Chandigarh 2004 has been released on 4th Oct. 2004 which portraits current state of environment in the U.T. Chandigarh and as well as facilitates the measurement of progress and to take remedial action to put the character and environs of Chandigarh in place

Budget of Plan Schemes for the Year (2006-07)

(Rs. In Lacs)

S.N.	Head/Item of the budget	Proposed expenditure during the year (2006-07)
	Ecology & Environment	
Env.1	Environment Research & Ecological	
a)	Direction & Administration	15.00
b)	Environment Education Training & Information	3.00
c)	Institutional Support & Public Participation	4.00
Env.2	Protection & Conservation of Resources	10.00
Env.3	Assistance to Chandigarh Pollution Control Committee	1.00
	Total Ecology & Environment	35.00
Env.4	Construction of Paryavaran Bhawan	50.00
	Total	50.00

* To be utilized by Engineering Department, U.T. Chandigarh.

The natural environment with its numerous living and non-living resources is man's most precious heritage. The basic goal of environmental conservation is the management of human use of these natural resources, so that they may yield the greatest sustainable benefits to present generation while maintaining their potential to meet the needs and aspirations of future generations.

Like other nations, India too bears the scars of damage done to its natural environment resulting in a wide array of environmental problems affecting the well-being of its citizens. While for developed countries, environmental problems are largely the by-products of affluence marked by resource wasteful life-styles, the stress on India's environmental resources come mainly from the pressures for satisfying the basic human needs of a large and growing population which has crossed 1 billion mark according to the Census of 2001.

Environmental problems in a developing country like India can be classified into two broad categories:

a) Those arising from conditions of poverty and under development.

b) Those arising as negative effects of the very process of development.

The first category has to do with the impact on the health and integrity of our natural resources (land, soil, water, forests, wildlife etc.) as a result of poverty and inadequate availability of the means to fulfil the basic human needs (food, fuel, shelter, employment, etc.) for a large section of our population. The second category has to do with the unintended side-effects of efforts to achieve rapid growth and development.

In this latter category would fall the distortions imposed on natural resources from not only poorly planned development projects and programmes, as well as from lack of attention to long-term interests by commercial and vested interests, but also the inherent and somewhat irreconcilable contradiction between development and ecological preservation.

Over the centuries, Indians have worshipped nature. Even in modern times, India can count itself as among the few countries of the world that have made specific reference in their Constitution to the need for promoting ecological security. The Indian Constitution lays down in the Directive Principles of State Policy the following duties for the State and the Citizen:

Article 48-A

"The State shall endeavour to protect and improve the environment and to safeguard the forests and wild life of the country".

Article 51-A

"It shall be the duty of every citizen of India-to protect and improve the natural environment including forests, lakes, rivers and wild life, and to have compassion for living creatures".

ECOLOGICAL SECURITY

One of the points in the Chief Minister's 15 point programme for realising the vision of making Tamil Nadu the best State in the country by providing opportunities for a healthy and productive life for all is ecological security. In the address of the State Finance Minister Budget session of the State legislative in March' 2002, it was emphasised that environment protection is the key to ensuring a healthy life for the people. Environmental problems are on the increase and are more pronounced in densely populated cities.

Exploding urban migration, as experienced in the last decade, is bound to widen the gap between demand and supply of infrastructural services such as energy, housing, transport, communication, water supply, sewerage, solid wastes disposal and recretation for communities. The mixing of the air with a high level of sulphur dioxide, carbon monoxide, oxides of nitrogen and suspended particulate matter by industries and vehicles is adding to air pollution. The area around Manali near Chennai, the belt of Vaniyambadi to Ranipet in Vellore District, textile and dyeing industries in and around Tiruppur & Erode towns and areas around tanneries in Vellore district are most prone to industrial pollution in Tamil Nadu.

The Tamil Nadu Pollution Control Board (TNPCB) and the Department of Environment of the Government of Tamil Nadu are entrusted with the task of monitoring pollution and taking efforts for abatement of pollution in the State. The Tamil Nadu Pollution Control Board enforces the provisions of Water (Prevention and Control of Pollution) Act, 1974 as amended in 1978 and 1988 and the Air (Prevention and Control of Pollution) Act, 1981 as amended in 1987 and the relevant provisions/ rules of the Environment (Protection) Act, 1986 for tackling pollution of water, air and land on account of causes such as industrial effluents, industrial gases, use of chemical pesticides, human negligence etc., Among its activities are (1) monitoring of treatment and disposal of effluents from the industries and local bodies and (2) survey of water resources in the State to assess their quality. The Department of Environment created in 1996 has the triple functions of implementation, co-ordination and monitoring of the State and Centrally Sponsored Schemes in such areas for pollution abatement.

Performance in Ninth Five Year Plan (1997-2002)

An outlay of Rs. 50 crores was provided for "Ecology and Environment" during the Ninth Five Year Plan period. The allocation to the Tamil Nadu Pollution Control Board was Rs. 15.87 crores for the implementation of thirteen schemes/ programmes and to the Department of Environment Rs. 34.13 crores for the implementation of six schemes / programmes out of the total sectoral allocation of Rs. 50 crores.

As against the Ninth Five Year Plan outlay Rs. 50 crores, the expenditure for the sector "Ecology and Environment "during the Ninth Five Year Plan period would be of the order of Rs. 11.69 crores representing approximately 24% of the Ninth Five Year Plan sectoral outlay of Rs. 50 crores.

The expenditure under Centrally Sponsored Schemes and schemes shared between State & Centre (Full Govt. Schemes) is estimated to be Rs. 11.81 crores and Rs. 17.49 crores respectively for the programmes of Ecology and Environment during the Ninth Five Year Plan period.

Tamil Nadu Pollution Control Board

The Tamil Nadu Pollution Control Board implemented the thirteen schemes / programmes envisaged during the Ninth Five Year Plan period with their own sources of funds. They are: (1) Hazardous Waste Management, (2) Strengthening of TNPCB Laboratories, (3) Noise Level Monitoring, (4) Vehicle Emission Monitoring, (5) Water Quality Monitoring of Non-Point Sources, (6) Establishment of Environmental Information System, (7) Coastal Monitoring Programme, (8) Monitoring of Industrial Estates, (9) Education and Training, (10) Air Quality Monitoring, (11) Emergency Response Centre, (12) Ozone Monitoring Cell and (13) Establishment of Environmental Management Cell for preparation of environmental maps, zoning atlas etc.

In addition to the above thirteen schemes/ programmes, the Tamil Nadu Pollution In Control Board took up the following Pollution control activities during the Ninth Five Year Plan period.

Major programme for development of green belt in municipal corporation areas to mitigate air and noise pollution in urban areas.

Detailed project reports for solid waste management at Udhagamandalam, Kodaikkanal, Palani, Thiruchendur, Kanyakumari, Rameswaram, Mamalapuram and Kutralam. Creation of awareness regarding the ecological hazards posed by plastics, through mobile exhibitions on eco-friendly alternatives to plastic materials and also through advertisement campaigns in public transport buses.

Establishment of a research fund and sanction of nine projects. Constitution of Public Hearing Panels as per the Environmental Impact Assessment Notification, 1994. Monitoring of 17 categories of highly polluting industries, specified by Government of India under which 202 units were identified. Initiation of necessary steps to implement the Recycled Plastics Manufacture and Usage Rules, 1999. Enforcement of the Bio Medical Waste (Management and Handling) Rules, 1998 as amended in 2000.

Towards setting up and operation of common effluent treatment plants in the State, the Government of Tamil Nadu sanctioned and released a subsidy amount of Rs. 1179 Lakhs and the Government of India released a subsidy amount of Rs. 1190 Lakhs during the Ninth Five Year Plan period.

Department of Environment

Five Old Towns: Pollution abatement works in river Cauvery at Bhavani, Pallipalayam, Erode, Komarapalayam and Tiruchy towns are being undertaken under the National River Conservation Programme during the plan period.

The programme became 100% Government of India supported programme since April 1997. The total estimated cost is Rs. 38.20 crores. The project period is 10 years and all the works will be completed by 2005. The expenditure during Ninth Five Year Plan period was Rs. 12.13 crores.

Seven New Towns: Government of India accorded administrative approval for the abatement of pollution in five towns on the banks of the river Cauvery viz Kumbokanam, Thanjavur, Karur, Mayiladuthurai and Tiruchy (additional areas) besides Madurai along Vaigai river and Tirunelveli along Tamirabarani river.

National Lake Conservation Programme: Government of India gave administrative approval for the implementation of the project for conservation and management of Kodaikanal and Ooty lakes at a total cost of Rs. 10.33 crores and Rs. 1.95 crores respectively.

Chennai City River Conservation Project: Under National River Action Plan, Government of India sanctioned a scheme at an estimated cost of Rs. 1700 crores with the participation of various Government agencies/ departments to clean Chennai City Waterways viz. Adyar, Cooum, Buckingham Canal, Otteri Nullah, Captain Cotton Canal and Mambalam Channel. Government of India grant for this project is Rs. 491.52 crores. The works, funded by Government of India, are executed by the Chennai Metropolitan Water Supply and Sewerage Board. The Public Works Department is involved in desilting of the waterways and sectioning, while the Tamil Nadu Slum Clearance Board is involved in rehabilitation and resettlement of 8000 slums on the fringes of the waterways. The Corporation of Chennai is involved in micro-drainage and solid waste management works. The Metro Water Board is also executing works to the tune of Rs. 228 crores from their own funds.

Coastal Zone Environment Management: The Government of India has constituted an authority called the 'State Coastal Zone Management Authority' with the Secretary to Government, Environment and Forests Department as Chairperson and the Director of Environment as Member-Secretary with five other official and non-official members.

Environmental Master Plans: The district profiles containing the details of data from primary and secondary sources have been prepared for all the districts and sent to the District Collectors and other user agencies. The preparation of status report on environment is nearing completion.

Environmental Education and Awareness: To enhance environmental awareness among students, over 1200 ecoclubs have been started in the schools. Eco clubs have been started in all the 30 districts by involving NGOs/ educational institutions. Financial assistance is also given for conducting workshops, training programmes to NGOs and educational institutions for conducting rallies on important days.

Strengthening the training facilities of the Department of Environment: Equipments such as OHP, LCD projector, colour TV, VCR etc. have been purchased to facilitate the conduct of communication and awareness programmes.

Tenth Five Year Plan

Objectives

The latest United Nation Report on environmental trends states that the expansion of ..ties, destruction of forests, erosion of fields and rising demand for water are likely to threaten human and ecological health in many countries for at least a generation. The report also observes that "Without the protection of environment, there can never be the kind of development needed to secure a fair deal for this or future generations".

Keeping this in view, the objectives of the Government of Tamil Nadu during the Tenth Five Year Plan period for the sector "Ecology and Environment" would be as follows:

- Protection of environment and reduction of wastes at source, besides continuing enforcement and monitoring activities.
- Conservation and enhancement of life support systems like land, water, forests, biodiversity, ocean and the atmosphere giving ecological security.

- Properly treating or converting the pollutants into useful products and thereby reducing their negative impact.
- Promotion of suitable measures and technologies for recycling/reusing.
- Emphasising the environmentally safe waste disposal options viz, composting of all organic wastes, common facilities for bio-medical wastes, etc. Maintaining essential ecological processes to ensure genetic diversity and prevent environmental degradation.
- Promoting the active participation of people at all levels-from the village panchayat to the State as a whole-in the preservation of the environment and promotion of sustainable development through awareness campaigns.

Besides the main two agencies viz., the Tamil Nadu Pollution Control Board and the Department of Environment, the various agencies/departments would also be addressing themselves to environmental concerns in the discharge of activities falling within their domain for the purpose of mainstreaming environmental consciousness and for tackling environmental issues on a holistic manner on all fronts in the State. The environmental concerns relating to various agencies/ departments are listed below:

Local Bodies

- Formulation and implementation of schemes for proper collection, conveyance, treatment and disposal of municipal sewage and solid wastes.
- Promotion and implementation of segregation of wastes at source, composting of wet wastes, recycling of dry recyclable and safe disposal of other wastes including toxics.
- Promotion of rainwater harvesting.
- Preparation of environmental management plans.
- Environmental awareness.

Tamil Nadu Water Supply and Drainage Board/ Chennai Metropolitan Water Supply and Sewerage Board

- Implementation of schemes for proper collection, conveyance, treatment and disposal of municipal sewage.
- Protection and monitoring of drinking water sources from impurities and contamination.
- Implementation of rain water harvesting systems.

Public Works Department

- Protection and monitoring of water resources including rivers, lakes and ground water.
- Sustainable development of major irrigation projects.
- Macro drainage network in river basins including proper de-sludging and desilting of river courses and water bodies.
- Implementation of statutory regulations on extraction of ground water.
- Action on rain water harvesting systems and enhancing the recharge capacity of ground water.

Industries Department/ SIPCOT/ SIDCO/ TIDCO

- Adoption of environmental features in industrial estates including green belt development, prohibition of disposal of toxic and non-biodegradable wastes including plastics.
- Proper collection, conveyance, treatment and disposal of sewage arising in the industrial estate.
- Implementation of environmental management plan in industrial estates.
- Adherence of environmental planning in new industrial estates.
- Action for evolving waste minimisation techniques and cleaner technologies for reduction/reuse/recycle of wastes in industries.

Tamil Nadu Electricity Board

- Action for utilisation of fly ash arising from thermal power plants.
- Improvement, operation and maintenance of air pollution control measures in thermal power plants to satisfy the standards prescribed.

Director of Town and Country Planning/ Chennai Metropolitan Development Authority

- Evolving environmental management plans for urban development.
- Review of master plan/development plan as per environmental considerations.
- Strict enforcement of land use classifications.

Forest Department

- Green belt development in the State.
- Implementation of urban afforestation programmes to control air and noise pollution.

Police and Transport Departments

- Action for control of vehicular pollution viz., emission and noise pollution.
- Proper traffic management especially at intersections and restriction of heavy vehicles in cities/towns during peak hours of the day based on the air quality monitoring data.
- Action for compliance of noise standards, especially during festival periods, celebrations, meetings, processions, etc.
- Curbing the use of air horns and cone speakers.

Food and Civil Supplies Department

- Check adulteration of fuel.
- Inspector of Factories

- Action for compliance of on site and off site emergency preparedness in major industries.
- Action for establishment of emergency response centres in industrial areas.

Health Department

- Monitoring of environmental epidemiology due to pollution.
- Conduct of pesticide epidemiological surveys.
- Action for establishing health care centres in major industries.

District Collectors

- District level review for effective implementation of environmental laws and programmes.
- Coordination for ensuring off site emergency preparedness.
- Implementation of provisions of the Public Liability Insurance Act, 1991. Ministry of Petroleum and Natural Gas
- Supply of cleaner fuels for automobiles

Major Thrust Areas / Strategies

Tamil Nadu Pollution Control Board

- Continuing enforcement of pollution control regulation and monitoring of industries.
- Capability building of local bodies to enable and empower them to monitor the environment and control pollution.
- Promotion of sewage treatment technologies in local bodies, especially decentralised treatment and low cost treatment such as root zone technology.
- Implementation of Municipal solid waste management rules, recently prescribed under Environment (Protection) Act,1986, focussing on source segregation and on composting of biodegradable wastes.

- Review of implementation status of hazardous waste management rules in the light of its recent amendments and establishment of storage, treatment and disposal facilities.
- Continuing the effort in regulating emission and effluents from large and medium scale industries.
- Sharp focus on small scale industries with emphasis on developing solutions for their problems.
- Focus on chemical-specific effluent/emission standards rather than only on lumped parameters such as chemical oxygen demand with measurement capabilities.
- Promotion of cleaner technologies starting from good house-keeping, recycle/reuse and raw material change towards waste minimisation.
- Promotion of life cycle analysis, environmental resource economics.
- Promotion of joint treatment (domestic with industrial waste water) in towns and cities, especially in places where new schemes are proposed.
- Focussing on drinking water source protection and on quality of drinking water supplies (catalytic role in the latter).
- Initiation of ground water and soil quality monitoring. Monitoring of pesticide residues in water and food. Maintaining coastline water.
- To improve data collection, storage, analysis and dissemination in respect of environmental quality.
- Ensuring that environmental concerns are adequately addressed by the development departments such as mining, industries, irrigation, agriculture, tourism, town and country planning, health, education, etc.
- Involving local bodies in environmental management by sensitising, training, promoting participatory decisions especially in locations of industries, planning regulations, regulation of certain activities such as congregations, etc.

- Creating awareness and promoting public participation in environmental issues, especially in municipal solid waste management.
- Synergy with universities and research institutes to promote relevant research studies, training and dissemination using such mechanisms as a standing committee of TNPCB, Department of Environment and Academic personnel.

Department of Environment

- Abatement of pollution in all the rivers in the State
- Coastal Legislation Zone management.
- Creating green lands in urban areas
- Municipal solid waste management
- Conservation of coastal eco-system and mangroves
- Development of eco-tourism
- Pollution free pilgrim centers
- Combating desertification Preparation of the Status of Environment report for Tamil Nadu
- Conservation of wetland eco-system
- Environment education and awareness in schools and colleges through NGO's and Eco-clubs.
- Synergy with Universities/Research Institutes for research, training and environmental education

PROGRAMMES/SCHEMES

Tamil Nadu Pollution Control Board

Schemes Carried over from Ninth Five Year Plan

1. *Hazardous Waste Management*: The Hazardous Wastes Management Cell in TNPCB will continue its functioning as per the recent amendment rules and enforce the concept of developing common facilities for treatment and disposal of hazardous wastes by industries/ agencies generating hazardous wastes. Strengthening of the Cell

will be continued with the assistance of Rs. 41.95 lakhs sanctioned by the Ministry of Environment and Forests, Government of India.

2. *Strengthening of TNPCB Laboratories*: The analytical capabilities of the TNPCB laboratories will be strengthened. Bio-monitoring in rivers Bhavani, Cauvery and Thamiraparani will be carried out through outside agencies (cost: Rs. 2 lakhs/ river/ year 3 rivers 5 years = Rs. 30 lakhs). It is also proposed to carry out a monitoring programme with the assistance of outside agencies at 5 places such as Thanjavur district, Udhagamandalam, Yercaud for non-point sources of pollution arising out of increased development activities in farming sector, which may cause discharge of persistent pollutants such as pesticides and nitrogen and phosphorous (Rs. 5 lakhs/ point 5 points = Rs. 25 lakhs). The monitoring of sources of production/ usage of ozone depleting substances will also be continued. The total cost will be Rs. 150 lakhs.
3. *Environmental Information System*: TNPCB has already started implementing the computerisation programme with its own funds and applying geographical information system with assistance from the Central Pollution Control Board. Networking of the data base with various offices of TNPCB and also CPCB is to be undertaken at a cost of Rs. 50 lakhs.
4. *Coastal Monitoring Programme*: TNPCB has already identified 7 hot spots in the coastal areas and is monitoring the hot spots. Additionally, 3 spots along with 7 already identified hotspots are proposed to be monitored with the assistance of outside agencies at a cost of Rs. 5 lakhs.
5. *Education and Training*: It is proposed to sensitise the local bodies and other Government departments so that they can integrate the environmental considerations in their development activities and also incorporate environmental concerns in their programmes/ policies.

Environmental awareness programmes are proposed to be continued for officials, non-governmental organisations, community based organisations, public, etc. It will be carried out at a cost of Rs. 250 lakhs.

6. *Air Quality Monitoring*: Establishment of continuous ambient air quality monitoring system at Mettur, and additional monitoring stations at Chennai (3 nos.) and Thoothukudi. TNPCB will persuade the industrial units in the clusters to bear the recurring cost of the system. Cost estimate: Rs. 40 lakhs per station 5 stations = Rs. 200 lakhs.
7. *Environmental Management*: Under the scheme, the following activities are proposed and the cost will be met from CPCB funding.
   I. Preparation of district wise environmental atlas for 8 districts viz, Chennai, Salem, Erode, Cuddalore, Karur, Dharmapuri, Tiruchirapalli and Dindigul at a cost of Rs. 60 lakhs.
   II. Industrial estate planning studies in 3 estates viz, (a) SIPCOT Industrial Estate, Sriperumpudur, (b) SIPCOT Industrial Estate, Phase-II, Hosur and (c) Petrochem Park, Tiruvallur District at a cost of Rs. 30 lakhs.
   III. Environmental management plan for 2 areas viz, Vellore and Tiruppur at a cost of Rs. 20 lakhs.

   The total cost estimate will thus be Rs. 110 lakhs.

New Schemes

Formation of New District Offices: TNPCB has established 25 district offices and is enforcing the pollution control legislations at the field level. In Vellore district, 2 district offices are functioning. In Coimbatore district also 2 district offices are functioning. Though the district offices cover all parts of the

State, it has been envisaged that in each revenue district a separate office must function and that certain districts have to be bifurcated for close monitoring of the sources of pollution and for protection of environment. It has been proposed to form 8 new district offices.

Cost estimate

(i) Non-recurring	Rs. 4.00 lakhs/district office
(ii) Recurring	Rs. 3.50 lakhs/district office
Total	Rs. 7.50 lakhs × 8 districts
	Rs. 60 lakhs.

Construction and Upgradation of TNPCB Laboratories: TNPCB has taken up a study on modernisation of its environmental laboratories with German experts, who are currently involved in a similar project for the Centre for Environmental Studies, Anna University. The preliminary investigation has been carried out. The cost for modernisation of the laboratories is estimated as follows:

Building	Rs. 200 lakhs.
Infrastructure	Rs. 150 lakhs.
Total	Rs. 350 lakhs.

Cleaner Technologies: Towards developing cleaner technologies, the following activities are proposed. The industrial contribution will be made a part of the scheme to ensure active participation of the beneficiaries. I. The cleaner technology options starting from good house-keeping, resource economics, recycle/reuse techniques, material management and process change are to be evolved for 5 categories of industries such as flour mill, electroplating, cashew nut processing, calcium-carbide and small dyeing at the cost of Rs. 15 lakhs. II. The life cycle analysis shall be carried out for 2 products at the cost Rs. 10 lakhs.

Total cost estimate: Rs. 25 lakhs.

Drinking Water Protection: Based on the inventorisation of major drinking water sources by Tamil Nadu Water Supply and Drainage Board, etc., monitoring of the quality of water sources will be carried out. Source protection plans are to be prepared for 5 important sources by external agencies involving the concerned departments.

Cost estimate: Rs. 4 lakhs per source 5 sources = Rs. 20 lakhs.

Solid Waste Management: TNPCB has already taken up the work on preparation of detailed project reports on proper solid waste management for 8 local bodies of important tourism centres. TNPCB is to continue sensitising the concerned officials and the general public on the need to segregate wastes at source, promote composting of wet wastes and recycling of recyclable dry wastes and minimize the amount of non-biodegradable wastes including plastics and domestic toxic wastes. For this purpose, a Municipal Solid Waste Management Cell is to be created in TNPCB to be incharge of municipal solid waste management and also bio-medical waste management by redeployment of staff.

Cost estimate: Rs. 6 lakhs per year for 5 years = Rs. 30 lakhs.

Vehicle Emission Monitoring: TNPCB has proposed to purchase 30 smoke meters and 5 carbon monoxide/hydrocarbon exhaust gas analysers with a view to integrate the vehicle emission monitoring with other departments and educational institutions.

Cost estimate: Rs. 65 lakhs.

Research and Development-Support to be extended for research and development activities especially in the area of pollution control technologies, waste minimisation options, assessment of quality criteria for drinking water, urban wet land conservation and development, etc., with availability options.

Cost estimate-Rs. 25 lakhs.

Environmental Education: According to the Constitution of India, it shall be the duty of every citizen to protect and improve the natural environment, including forests, lakes, rivers and wildlife and to have compassion for living creatures.

Environmental education has become an inevitable tool in creating awareness on protection of environment. The major goals are (i) to foster clear awareness of, and concern about economic, social, political and ecological interdependence in urban and rural areas, (ii) to provide every person with opportunities to acquire the knowledge, values, attitudes, commitment and skills needed to protect and improve the environment and (iii) to create new patterns of behavior of individuals, groups and society as a whole, towards the environment. Curriculum development and faculty training in universities and colleges are to be carried out.

Cost estimate-Rs. 50 lakhs.

Environmental Impact Assessment: Conducting regional environmental impact assessment studies at 10 places such as Cuddalore, Mettur, Ranipet, Manali, Tiruppur and Thoothukudi. Such studies will also be conducted for a cluster of lime quarries in Tiruchirapalli and Thoothukudi districts. The study will assist in exploration of quarry activities in a sustainable manner in those areas. Funding by industrial sector will be explored.

Cost estimate-Rs. 10 lakhs per area 10 areas = Rs. 100 lakhs.

Environmental Epidemiology Studies: Since the Stockholm conference, both the need for environmental protection, and awareness of the links between environmental protection and human health, have grown considerably. Studies on environmental epidemiology with specific reference to local conditions are to be conducted with assistance from outside agencies. Health effects due to pollutants, human health risk assessment, exposure to indoor air pollutants, etc. are the various components of study.

Cost estimate Rs. 50 lakhs.

Prohibition on use of Plastics

'Omnipresent, the plastic carry bag has strewn itself everywhere. In gardens, parks, drains, garbage dumps, on branches of trees and even in bird nests, it can be found to exist, propagating almost like a life form. Light

and flimsy, the ubiquitous carry bag floats around in a light breeze, finding new places to litter and inhabit forever.' (an environmentalist, 1996)

Recognizing the damage which the use of plastic, which is non-biodegradable, causes to the environment, a Bill was moved in the Tamil Nadu Legislative Assembly in May 2002 to prohibit the sale, storage, transport and use of non-reusable articles made of, or containing, plastic in the State. This Bill has been referred to a Select Committee.

Environmental Health Cell: A key reason for putting in place laws and regulations pertaining to pollution and environment and establishing regulatory bodies to monitor and ensure the observance of such laws and regulations is the concern over the health implications of pollution and environmental degradation. While the health department is mandated to monitor morbidity and mortality parameters in the population, there is no specific focus on the impact of pollution and environmental degradation. Under the circumstances, to monitor the health indicators either in and around industrial sites or in municipalities and other urban settlements, it would be highly worthwhile to set up an environmental health unit in TNPCB. The objective is to liaise and coordinate with the Health department, Inspectorate of Factories, etc., Cost estimate

(i)Non-recurring	Rs. 25 lakhs.
(ii) Recurring	Rs. 35 lakhs for 5 years.
Total	Rs. 60 lakhs.

Institute of Environmental Management: In environmental protection, every nation has an important role to play as a follow up to the Earth summit, 1992 and implementation of Agenda 21. The effort should be taken in a comprehensive manner so that both environment and development concerns are afforded the highest priority, as the overall objective of Agenda 21 is to integrate environment and development issues at all levels of decision making, viz. local, regional and international. To achieve the objectives of sustainable development, evaluation of training needs, promulgation of environmental policy issues, etc. and also to function as an environmental data centre, there is an urgent need to establish an Institute of Environmental Management at national level. Necessary contributions, if any from the State Government, will be made in the form of land, etc., as and when concurrence comes from the Government of India. Cost estimate: Rs. 4000 lakhs. The source of funding for each schemes / projects as well as the phasing of outlay for the Tenth Five Year Plan period for the Tamil Nadu Pollution Control Board are given in the Table.

DEPARTMENT OF ENVIRONMENT

Schemes Carried over from Ninth Five Year Plan

Providing Green Cover for Clean Air: Urban forestry is rather a neglected area in Tamil Nadu. An attempt was made in the Ninth Plan to green the landscape of urban areas at a cost of Rs. 4 crores. Plants have the ability to remove significant quantities of pollutants from the air without sustaining serious growth retardation. Vegetation in any form will be a scavenger or a sink of many air-borne substances like SO_2, NOX, Cl_2, etc. Educational institutions, industrial establishments etc. having sufficient space may raise suitable tree species and maintain them. For undertaking this activity, they may be given annual awards with trophies and certificates. Identified institutions will be supplied suitable seedlings. This scheme requires an outlay of Rs. 50 lakhs for Chennai and Rs. 20 lakhs per annum for other 5 corporations. For 5 years an outlay of Rs. 7.50 crores for 6 Corporations in the State is proposed.

New Schemes

Conservation of Coastal Eco-System: Coastal area forms a highly complex and dynamic eco-system, sensitive to developmental pressures. With the growth of cities, towns, ports, industrial establishments and other developments, great expanse of the coasts have been altered to such an extent that they now hardly resemble the original form; degradation and multi-utilisation of beaches are affecting the aesthetic and environmental values with significant economic and recreational loss. These could be avoided through prudent coastal development and management based on assessment of ecological values and potential damage from coastal developments. An integrated Coastal Zone Management Plan involving various agencies and stakeholders will be prepared for the Tamil Nadu coast during the Tenth Plan engaging suitable consultants at a cost of Rs.1 crore. A provision of Rs. 108.48 crores is made in the Plan for fixing the High Tide Line (HTL) in respect of important towns, urban areas such as Cuddalore, Thoothukudi, Nagapattinam and Kanniyakumari.

Lana Management

Combatting Desertification: The phenomenon of desertification is very critical throughout the world as evident from assessment and various reports of UNEP. Desertification has taken place over centuries. It has been accelerated by human activities to secure a livelihood under ecologically fragile conditions. This programme envisages coordinated implementation along with various departments such as Department of Agriculture, Horticulture, Forests, Agricultural Engineering and Public Works Department. This department will take up the following programmes:

- Identification of areas requiring drought proofing in the State
- Socio-economic condition of the area
- Identifying and forming of environment friendly self-help groups for carrying out planting/water harvesting, etc.

An integrated project for Rs. 5.50 crores will be prepared during Tenth Plan and sent to Ministry of Environment and Forests, Government of India for funding.

Scheme to Reverse Loss of Ecology: Cultivable land has been affected seriously due to industrial pollution in 6 districts. As per the Supreme Court's direction, Loss of Ecology Authority has estimated the loss to the tune of Rs. 27 crores in Vellore

District. Besides pollution abatement strategy, the reversal of land degradation will have to be undertaken. Loss of Ecology Authority has also obtained certain proposals and submitted to the State Government. It is proposed to undertake special programme for the reversal of land degradation. This will be partly funded by the industries responsible for the loss of ecology besides availing the pooled state and central funds. A provision of Rs. 10 crores has been made during Tenth Five Year Plan for undertaking these works.

Preparation of the Status of Environment Report: This is an important aspect of the Tenth Plan period. It will be useful to all the departments of the State in implementing various schemes. The financial outlay required is Rs. 15 lakhs of which Government of India's grant will be Rs. 10 lakhs.

Conservation of Wetland Eco-System: Conservation of wetlands especially around the cities and towns is critical for recharge of ground water and aesthetic amelioration. At present, these wetlands and swamps, which are ecologically sensitive, are highly threatened. Quite often these wetlands are converted into dumping yard for garbage or polluted by sewage and other effluents.

They are encroached and built up in several cases. Hence, conscious efforts are needed to identify these threatened wetlands and conduct scientific study and action plan for their eco restoration. Most critically threatened wetlands could be declared as needing protection under the RAMSAR convention. An amount of Rs. 20 lakhs is provided in the Tenth Plan for undertaking survey of wetlands, identification of threats and action plan for eco-restoration.

Eco-cities Programme

(a) The concept of eco-city is of a place where environment friendly practices prevail in all public services including management of municipal solid and liquid wastes, bio-medical wastes, hazardous wastes, protection of heritage structure monuments and sensitive ecosystems, increasing green cover through urban forestry besides education and involvement of people in environment protection. Environment Management Plan will be prepared for Madurai, Rameswaram, Mahabalipuram, Ooty, Tiruvannamalai and Courtallam during the Tenth Plan.

(b) Besides preparing Environment Management Plans, selected Municipalities such as Ambattur/ Poonamalle, Town Panchayats such as Pammal and Village Panchayats such as Madipakkam around Chennai will also be showcased as model towns and cities.

(c) Solid Waste Management-The burden of excessive human claims on the earth's biological system is aggravated by yet another human excess-the generation of waste. Due to accelerated human activity, the quantum of generation of solid wastes in the urban agglomeration is increasing manifold.

In urban and semi-urban areas, the solid waste contains both bio-degradable part and non-biodegradable wastes. The management of solid wastes in corporations and municipalities has to be given a new thrust during the Tenth Five Year Plan.

One of the basic methods in the management of solid wastes is segregation at source.

(1) Every street or colony may segregate waste using two bins-one for collection of bio-degradables and another for non-bio degradables.

(2) Formation of aerobic composting/ vermi composting pits for the disposal of biodegradable material will be encouraged.

(3) Hand made paper units/plastic recycling units will be established for non-bio degradables through NGOs/ industry wherever feasible.

The above three activities will be implemented in a phased manner in all municipalities taking a certain number of municipalities every year. The financial outlay for each municipality will be Rs. 10 lakhs. This will be a one-time expenditure. Thereafter, every year, the municipalities themselves will have to manage the solid waste management system and generate enough wealth from waste to manage and sustain the system.

During the Tenth Plan, the programme will be taken up in 6 municipalities by inviting suitable proposals and the best proposal will be selected. Participation of NGOs and private sector will be encouraged to manage the solid waste in a scientific manner. For training programmes and supply of requisite publication materials, etc. Rs. 10 lakhs is provided at the rate of Rs. 2 lakhs per year in the Tenth Plan. Total financial commitment will be Rs. 70 lakhs.

		Rs. in lakhs
a)	Environment Management Plan for 6 towns	
	@ Rs. 2.50 lakhs	15
	Action Programme on pilot scale	
	@Rs. 15.00 lakhs/town for 6 towns	90
b)	EMP for selected towns around Chennai	
	@ Rs. 2.50 lakhs/town	10
	Action Programme @ Rs. 10.00 lakhs/town	40
c)	Solid Waste Management in six selected municipalities,	
	including awareness programmes	70
	Total	225

Bio-diversity Conservation Network in Tamil Nadu: Tamil Nadu has a rich biodiversity. Efforts to conserve biodiversity are being made by the departments of Forests and Agriculture, involving NGOs in the process. It is proposed to set up a network on biodiversity conservation involving diverse organisations and giving assistance for promoting biodiversity conservation.

This programme will involve in-situ conservation, ex-situ conservation, documentation and networking. Besides conventional state level seminar every year, modern methods of internet video conferencing will be used for network and bringing together the efforts being made by various agencies for biodiversity conservation. We will seek funds for this programme from Ministry of Environment and Forests. The project will involve the following.

		Rs. in Lakhs
a)	Conservation measures	30
b)	Networking	10
c)	Seminar, Documentation (Soft & hard copies)	10
	Total	50

Environmental Research and Development: Specific need based environmental research will be promoted in the areas of low-cost waste water treatment, solid waste management, problem soil treatment under combating desertification programme, River/lake water quality monitoring package for recovery and recycling urban forestry and aesthetic plantation to minimise air and noise pollution in cities.

A provision of Rs. 120 lakhs is made in the Tenth Plan for specific research to fill up the research gaps in knowledge and develop new methods and technology. Research capability of various universities such as Anna, Madras, Tamil Nadu Agricultural, Madurai Kamaraj, Bharathiar and Bharathidasan etc., will be made use of. The service of NGOs will also be utilised effectively.

Agency for Environmental Management: In order to coordinate with various implementing agencies and the Government of India for flow of funds, approval of DPRs and effective implementation, supervision and monitoring of the river cleaning and lake conservation projects, it is proposed to create two autonomous agencies at Chennai and Tiruchy on the lines of DRDA.

This agency will be an autonomous agency like DRDA and will be entrusted with works such as river cleaning, lake conservation, environmental awareness and coordinate with NRCD, TWAD Board, Metro Water and other Municipalities.

Jurisdiction of the Agency-Chennai

1) Chennai City River Cleaning Programme Chennai
2) National River Action Plan Works at Erode and Namakkal Districts
3) National Lake Conservation Programme-Ooty Lake

Jurisdiction of the Agency-Tiruchy

1) National River Action Plan at Tiruchy
2) National River Action Plan at 7 towns on the Cauvery rivers
3) National River Action Plan at Madurai and Tirunelveli
4) National River Action Plan at Coastal Towns of Rameswaram and Tiruchendur
5) National Lake Conservation Programme at Kodaikanal.

Therefore, it has been proposed to create two agencies one at Chennai and another at Trichy, with jurisdiction as detailed above for effective execution and monitoring of various schemes. Out of total expenditure of Rs. 63 lakhs per annum, the recurring expenditure will be Rs. 46 lakhs per annum for both the agencies. The expenditure is proposed to be met from 8% centage charges permitted by Government of India towards staff cost. An outlay of Rs. 250 lakhs is provided in the Tenth Plan.

Environment Education and Awareness-This is an ongoing programme. Eco-clubs in schools in the State will be continued. Totally, 1200 schools spread over 30 districts will be used for spreading environment messages throughout the State. Besides, National Green Corps schemes of Government of India will be formed in 2900 schools at the rate of 100 schools per district.

Each of the 1200 eco clubs will be provided Rs. 1000/-for undertaking environmental awareness in their campus and surrounding inhabitation. For this purpose, training in solid waste management, rainwater harvesting, etc., would be given to them.

a)	Enhancement of environmental awareness through eco-clubs 30 × 0.50 × 5	Rs. 75 lakhs.
b)	Special awareness programme in selected towns 5 × 5	Rs. 25 lakhs.
c)	National Green Corps in 2900 schools Rs. 1000/- per school per year for 5 years	Rs. 150 lakhs.
	Total	Rs. 250 lakhs

The source of funding for each schemes / projects as well as the phasing of outlay for the Tenth Five Year Plan period for the Department of Environment are given in the Table at the end of this section.

Outlay

Ecology and Environment sector will thus have a total outlay of Rs. 197.20 crores during the Tenth Five Year Plan period, out of which, the state budget outlay will be Rs. 113.05 crores. The details are given below. The sources of funding for the Tamil Nadu Pollution Control Board and the Department of Environment are also indicated.

2

REVIEWS ECOLOGY AND ENVIRONMENT DEPARTMENT

IMPLEMENTATION OF THE AIR (PREVENTION AND CONTROL OF POLLUTION) ACT

Implementation of The Air (Prevention and Control of Pollution) Act, 1981, The Environment (Protection) Act, 1986 and rules framed thereunder and Hazardous, Bio-medical and Municipal Solid Wastes Management Rules, by State Government through Karnataka State Pollution Control Board (Board) and Commissioner of Transport and other authorities was inadequate and ineffective. Only 0.13 lakh out of total 2.70 lakh industrial units are under the purview of these Acts and Rules. Ambient Air Quality Monitoring Stations (AAQMS) were confined to only two cities (Bangalore and Mysore). Board had not established air quality monitoring stations in other cities though funds were available.

Due to lack of adequate monitoring stations, analysis of air samples by the Regional Offices was inadequate. Analysis of emission from industries was only marginal. Particles generated by stone crushing units were not analysed at all. Automobiles contributed to air pollution on a large scale (2357.92 MT per day in Bangalore city and 6116.49 MT for the State). The Commissioner of Transport, however, did not monitor implementation of this component. In 3 districts, no inspection of vehicles was carried out by Regional Transport Officers (RTO). Board had no information about the quantity of bio-medical and hazardous waste generated and its disposal by 50 per cent of health care establishments and industries. In view of these deficiencies, the Board's machinery was largely non-functional.

Board had not identified/covered 95 per cent of industries registered with Department of Industries and Commerce. Out of 0.13 lakh industries identified, 7420 were in operation as of March 2001. Of these, 2623 and 605 industries respectively were working without consent and Air Pollution Control Systems in violation of Acts/Rules and 2308 had not even applied for consent. Diesel Generator sets and Stone crushers had not been brought under the purview of Acts/Rules.

INTRODUCTION

Presence of pollutants in air beyond certain limits, discharged by industries and heating process etc., cause environmental pollution and adversely affect living organisms and natural resources. Government of India enacted following Acts and Rules for prevention and control of air and other pollution of environment:

(i) The Air (Prevention and Control of Pollution) Act, 1981 (Act)

(ii) The Environment (Protection) Act, 1986 and Rules 1983 (Acts/Rules)

(iii) The Hazardous and Bio-medical Wastes (Handling and Management) Rules (Rules), 1989

(iv) Municipal Solid Wastes (Management and Handling) Rules, 1999. In June 1992, Government of India framed a policy statement for abatement of pollution.

Government of Karnataka framed the Air (Prevention and Control of Pollution) Rules in 1983. These Acts and Rules prescribed, inter alia, the following measures for prevention, control and abatement of pollution:

(i) Coverage and identification of industries to bring them within the purview of Act and Rules.

(ii) Setting standards for ambient air quality as well as standards for emission of air pollutants by industries and automobiles.

(iii) Establishing laboratories.

(iv) Prosecution of such industries/persons not complying with the prescribed standards of emission.

Organisational Set Up

The Principal Secretary, Forest, Environment and Ecology was responsible for implementation of Acts/Rules and monitoring of various programmes for prevention and control of Air Pollution. While Karnataka State Pollution Control Board (Board) was the main agency for implementation of Acts and Programmes, other Departments (Police, Transport and Forest) supplemented the efforts of the Board.

Audit Coverage

Implementation of the Air Act and Rules was reviewed through test-check of records in Regional and Divisional Offices of the Board, Regional Transport Office, District Industries Centres, Zilla Panchayats, Municipalities and Corporations in six districts ?. Besides, details were collected from Secretary, Ecology and Environment Department, Commissioners for Transport and Industries and Commerce. The important audit findings are discussed below:

Budget Provision and Expenditure

As against budget provision of Rs.37.70 ? crore, expenditure was Rs.10.54 crore on various programmes of Department of Ecology and Environment which included expenditure of Rs.1.84 crore for Prevention and Control of Air Pollution from 1996-97 to 2000-01. Out of Rs.1.84 crore, Rs.37.31 lakh released to the Board during March 1999 for purchase of Ambient Air Quality Monitoring equipment remained unutilised as of March 2001. No reasons had been furnished for non-utilisation of these amounts. Board spent Rs.44.30 crore out of total receipts of Rs.73.73 crore (Central grant Rs.0.78 crore, State Government grant Rs.7.39 crore, Central Pollution Control Board (CPCB) grant Rs.0.79 crore, Board receipts Rs.63.03 crore and opening balance of Rs.1.74 crore as on 31 March 1996). Board had accumulated investment of Rs.37.23 crore as on 31 March 2001.

Poor Identification and Coverage of Polluting Industries

According to Section 17 of the Act, Board was to plan comprehensive programmes for prevention and control of air pollution, conduct inspection of air pollution control areas, collect and disseminate information on industries causing air pollution. Further, no person could operate any industrial plant in an air pollution control area without obtaining consent from Board as per Section 21 of the Act. This implied that Board was to conduct periodical survey in order to identify industries causing air pollution and bring them under consent regime. Board had neither conducted survey nor prepared any comprehensive programmes for coverage of air polluting industries.

However, on the basis of information furnished by the Department of Industries and Commerce, Chief Inspector of Factories and Boilers and Inspections, Board identified 0.13 lakh (4126 industries during 2000-2001) out of 2.70 lakh industries registered (March 2001) with Department of Industries and Commerce.

Out of 395 industries cleared by High Level Committee (34) and State Level Single Window Agency (361), 114 industries (9 with investment of more than Rs.50 crore, 105 with investment of more than Rs.1 crore but less than Rs.50 crore) had been established during 1998-2001. Board had no information regarding number of industries working without consent and Air Pollution Control Systems (APCS) even out of these 114 industries. Details regarding number of industries cleared by District Level Single Window Agency working without consent and APCS were also not available with the Board. However, the position regarding industries working without consent and APCS out of 0.13 lakh industries (Red-4881, Orange-2241, Green-5812) were as follows:

Of 2623 industrial units which had not obtained consent, 2308 units had not even applied for consent and continued to operate in contravention of provisions of Section 21 of the Act. Details regarding the period from which these had been working without consent and adoption of APCS were also not available with the Board. Thus, monitoring of the defaulting units was ineffective.

Even though Board had left out 95 per cent of industries and not categorised these as red, orange and green, it did not indicate any specific steps to cover huge shortfall in identification of industries but merely stated (June 2001) that action would be taken to cover all air polluting industries.

Regional Officers (RO) (Raichur, Mangalore and Mysore) categorised 5 industries engaged in manufacture of industrial gases, liquor, oxygen and ice and discharging pollutants into atmosphere as non-polluting on the basis of their own inspection. However, audit scrutiny of inspection report and other records regarding manufacturing process revealed that these industries were actually discharging pollutants into atmosphere. Reasons for such wrong classification was not evident. At the instance of Audit, ROs agreed to bring these industries under the purview of the Acts. Thus, the categorisation of polluting industries needs scrutiny. Board needs to examine the matter to fix accountability. As of March 2001, though 286 foundries and 1596 brick kilns, which were polluting, were registered with the Department of Industries and Commerce, the Board had not taken any steps to bring these units under consent and ensure adoption of APCS as prescribed in Acts/Rules.

National Ambient Air Quality Monitoring System

(a) Pollution Status

With a view to evaluating air quality and assess pollution status, CPCB introduced in 1984-85 'National Ambient Air Quality Monitoring System' (NAAQMS). This was also implemented by the Board in addition to its own programme. The system envisaged establishing monitoring stations and sample analysis for every eight hours at each station and assessment of presence of primary pollutants (Suspended Particulate Matter-SPM, Respirable Suspended Particulate Matter-RSPM, Sulphur di-oxide-SO 2, Oxides of Nitrogen-NO X) as against standards prescribed under Act/Rules. In respect of other two pollutants (Oxides of Nitrogen and Sulphur Dioxide) the prescribed standards exceeded by 3 to 81 per cent during 5 to 13 months during 1997-98 to 2000-01. Ambient air quality had not been monitored for Lead and Carbon Monoxide. According to results obtained by the Principal, Bapuji Institute of Engineering and Technology, Davanagere who conducted ambient air quality test (SPM only) in 6 locations in Davanagere city and Harihar town from 1998-99 to 2000-01, the presence of SPM ranged between 661 to 739 (3 to 4 times above limit) in one location and 236 to 646 in other locations.

Board had not established any additional AAQMS after 1996-97 as proposed in the Annual plans though there was no shortage of funds

(b) Shortfall in Ambient air Quality Tests

As prescribed in Schedule VII to Rule 3B of Environment Rules, 1986, ambient air quality test was to be monitored every eight hours for each station twice a week. According to guidelines/norms prescribed by CPCB and adopted by the Board, number of samples to be analysed for 3 to 4 pollutants (4 from 1999-2000) each year for seven stations (five under NAAQM and two under Board Programme) established at different locations in Bangalore and Mysore and achievement thereagainst were as under:

Year	Number of samples to be analysed	Number of samples actually analysed	Shortfall
1997-98	10920	4240	6680 (61%)
1998-99	10920	6436	4484 (41%)
1999-2000	11232	7992	3240 (29%)

Thus, there was 61 to 29 per cent shortfall in sample analysis for ambient air quality for which no reasons were assigned.

In respect of other cities/areas where AAQMS had not been established, Regional Officers who had been provided with necessary equipment and technical/scientific staff were to collect and analyse samples for monitoring air quality.

However, the Board had not prescribed any target for sample analysis. In five test-checked districts (Davanagere, Raichur, Mangalore, Hassan, Dharwad), number of samples analysed ranged from 23 to 51 each year and total of 659 samples were analysed during 1999-2001. In the absence of targets, the adequacy or otherwise of sample analysis could not be verified in audit.

(c) Non-functional Ambient Air Quality Monitoring Stations

State Government released (March 1997) Rs.30 lakh to the Principal Investigator, Centre for Atmospheric Sciences, Indian Institute of Science, Bangalore for design, fabrication and setting up of five AAQMS and their maintenance for one year. These stations established in December 1998 at different places in Bangalore city stopped working frequently and became defunct from June 1999 even before expiry of maintenance period due to malfunctioning of certain components. Though these components were reportedly (February 2000) replaced, the equipment remained non-functional. Thus, expenditure of Rs.30 lakh on these stations had been unfruitful.

Shortfall in Analysis of Air Emission

(a) Emission from Industries

According to provisions of Section 22 of the Act, no industrial or processing unit or person can discharge into air, emissions containing environmental pollutants in excess of the prescribed standards. Board was to ensure compliance with this provision by drawing samples of emission and analysing the same. According to norms prescribed by the Board, details regarding number of samples to be drawn and analysed on the basis of number of industries in operation and achievement thereagainst were as follows:

Year	Number of samples to be drawn	Number of samples actually drawn	Shortfall	%
1996-97	7609	308	7301	96
1997-98	7371	502	6869	93
1998-99	8380	404	7976	95
1999-2000	10062	1157	8905	89
2000-01	10465	1097	9368	90

Member Secretary of the Board stated (May 2000) that necessary action would be initiated to avoid such shortfall. Out of 3468 emission samples collected and analysed, details regarding number of samples in respect of which prescribed standards were not fulfilled had not been maintained by the Board.

However, in 39 out of 107 emission samples analysed by Central laboratory from 1998-99 to 2000-2001, pollutant (SPM) exceeded the prescribed standard. In three out of 6 test-checked districts, number of samples (emission from industries) drawn were meagre each year ranging from 1 to 5 during 1997-2001 as against 4752 samples required to be analysed as per norms. Stack samples were not at all analysed for four, three and two years in Davanagere, Hassan and Dharwad districts. The ROs attributed the shortfall in sample analysis to inadequate staff.

The reply was not tenable as shortfall in sample analysis was almost hundred per cent as against deficiency of 39 per cent of sanctioned strength in scientific and technical cadres (sanctioned-36, working-22). Out of total 12 samples drawn in these three districts, in respect of three samples, pollutants exceeded the standards prescribed.

Further, scrutiny of Register of Sample Analysis revealed that emission samples had been analysed generally for SPM only and samples had not been analysed for other pollutants (SO_2 and NO_2). Large shortfall in sample analysis and failure to analyse samples for all types of pollutants rendered implementation of the Act ineffective.

(b) Particles Generated by Stone Crushers

Board was to analyse samples for SPM in respect of particles generated by stone crushers twice a month throughout the year for each stone crusher, drawing samples at a distance of 40 metres from an isolated unit as well as from a unit located in a cluster in order to ensure that presence of SPM was within the standards prescribed in Schedule 1 to Rule 3 of Environmental Rules.

Though 916 stone crushers were working in the state, particles generated by stone crushers had not been monitored at all. The huge shortfall in analysis of air samples (61 to 29 per cent) and emission from industries (96 to 89 per cent) and total absence of analysis in respect of stone crushers was not justified as Board had 74 per cent of sanctioned strength in technical and scientific staff and appropriate equipment.

Evidently, there is need for examining the priority set by the Board for these areas.

(c) Emission due to Fuel Wood Consumption

According to standards prescribed (February1989) by the Board, when one MT of fuelwood is burnt, it causes emission of air pollutants (SPM) of 6.5 kg. During 1997-98 to 2000-01, 7.78 lakh cubic metres (2.73 lakh MTs) of fuelwood had been sold by Forest Department. According to standards prescribed by the Board, 1771.27 MTs of SPM had been emitted into air. The consumption of fuelwood for domestic purpose could be reduced to the extent of 40 per cent (as per circular issued in July 1994 by Rural Development and Panchayat Raj Department) by using to full extent the improved chullahs distributed under 'National Programme of Improved Chullahs' (NPIC) implemented by Zilla Panchayats (ZP)/Taluk Panchayats (TP).

However, during 1996 to 2001 as against targeted distribution of 1.26 lakh chullahs under the scheme, only 0.42 lakh chullahs (33 per cent) were distributed in five test-checked districts. Failure of ZP/TP to achieve the prescribed target affected air pollution adversely.

(d) Non-monitoring of Emission from Diesel Generator sets

As of March 2001, 19764 Diesel Generator (DG) sets had been registered with Electrical Inspectorate. The Board had not brought under the purview of Acts/Rules those DG sets installed in other than industries. In respect of DG sets installed in industries, the Board stated (September 2001) that as no limit was fixed for sulphur dioxide, emission from DG sets had not been monitored. The reply was not tenable as it was the responsibility of the Board to fix the limit as prescribed in Section 17 of the Act and arrange for their monitoring due to the potential of pollution from these industries.

Inspection

According to Section 24 of the Act, the Board was vested with the power to conduct inspection of industrial or processing units etc., to ascertain compliance of the provisions in Acts/ Rules, examining/testing pollution control equipment. According to norms prescribed (February 1996) by the Board, frequency of inspection was as follows:

Red #	Orange#	Green#	Medium
Once in 3 months	Once in six months	Once in a year	Small
Once in 6 months	Once in a year	Once in two years	Large
Once in every month			

Details regarding categorisation of industries as red, orange and green were available from 1999-2000. Targeted inspections and achievements as per the earlier norm of 60 inspections per Regional Office during 1996-97 to 1998-99 and as per the revised norms from 1999-2001 were as follows:

Year	No. of inspections to be conducted	No. of samples conducted	Shortfall	%
1996-97	10800	1922	8878	82
1997-98	10800	8100	2700	25
1998-99	10800	9175	1625	15
1999-2000	19372	10579	8793	45
2000-2001	26481	12123	14358	54

For 2.73 lakh MT of fuelwood purchased and used by public and Tile factories, SPM emitted was 1771.27 MT.

Though number of inspections increased, the units not inspected also increased sharply. This indicated that the Board needed to reschedule its inspections appropriately. Member Secretary stated (May 2001) that shortfall in inspection was due to priority given for inspection of highly polluting industries. The reply was not tenable as Board had not maintained separate details regarding number of inspections of industries categorised as Red.

However, as against 4596 and 6624 inspections to be conducted as per norms in respect of 383 and 552 Large Red category of industries, 1824 and 1753 inspections were conducted in respect of all large industries during 1999-2000 and 2000-2001 respectively. Even then, there was shortfall of 60 and 74 per cent during 1999-2000 and 2000-01 respectively. Due to inadequate inspection, larger number of industries operating without consent and without adopting APCS were not detected.

Failure to Monitor Noise Pollution

State Government/Board did not monitor noise pollution on regular basis except on complaints. Details regarding number of complaints attended, number of cases where noise pollution exceeded prescribed standard vide Rule 3 of Environmental Rules, 1986 and action initiated on such cases were not available with the Board. Even in test-checked districts, measurements taken for monitoring of noise pollution were insignificant (11 measurements by 3 Regional Offices and 90 measurements by 2 Regional Offices) from 1996-97 to 2000-01. State Government had not categorised areas into industrial, residential or silence zone/areas (area within less than 100 sq. metres around hospitals, educational institutions and courts) as required under Noise Pollution (Prevention and Control) Rules, 2000.

POOR MANAGEMENT OF BIO-MEDICAL WASTE

The Bio-medical Waste (Management and Handling) Rules were introduced in 1998. The Board being the prescribed authority for enforcement of provisions of these Rules was to identify establishments generating bio-medical waste, issue authorisation and regulate storage and disposal of waste in accordance with relevant rules.

As of March 2001, Board identified 3345 Health Care Establishments (HCE) (Government HCE-2449 and private HCE-896) producing bio-medical waste. Of these, 73 HCEs (which had a bed strength of 31838) were to comply with these rules by December 2000.

However, only 40 of these HCEs (Government HCE-20 and private HCE-20) obtained authorisation while 9 HCEs (Government HCE-4 and private HCE-5) had not even applied for authorisation and applications in respect of 24 HCE (Government-18, Private-6) were under process. In respect of remaining 3272 HCEs each with bed strength ranging from 1 to 200, 716 obtained authorisation.

Out of 40 HCEs, 10 HCEs (Government HCE-9 and private HCE-1) with a total bed strength of 6568 producing 3285 Kgs of bio-medical waste per day had not established waste treatment facility such as incineration, autoclaving etc. and bio-medical waste generated by them was disposed in municipal dumping sites and open grounds and mixed with municipal waste.

Another 13 HCEs, each with a bed strength of less than 200, were disposing of bio-medical waste into municipal waste. Out of 73 HCEs, only two (Kasturba Medical Hospital, Manipal and KLE Society Medical College Hospital, Belgaum) HCEs had established effluent treatment plant for treatment of liquid bio-medical waste. The details regarding the quantity of bio-medical waste generated and its disposal by 33 HCEs (Government-23 and private-10) (each with a bed strength of more than 200) which are working without authorisation were not available with the Board. The HCEs were to furnish to the Board annual returns indicating quantity of bio-medical waste generated, treatment and its disposal.

However, so far 150 HCEs had furnished annual returns as of March 2001. The Board stated (September 2001) that show cause notices were issued to defaulting HCEs.

Municipalities/Corporations were to grant license while Director of Health and Family Welfare grants recognition to the private HCEs. These authorities also had no information regarding the quantity of bio-medical waste produced and its disposal by 11 private HCEs working without authorisation. Thus, there was need for co-ordination among these agencies.

Hazardous Wastes Management

The Hazardous Wastes (Management and Handling) Rules, 1989 were introduced in July 1989. The Board being the prescribed authority for enforcement of these rules had identified 944 industries (dyes, chemical, petrochemical, natural gas etc.) as of March 2001.

According to Rule 5(2), these industries were to apply and obtain authorisation within 6 months from the date of commencement of these Rules. As of March 2001, out of 944 industries identified 591 only had applied for authorisation and 478 had been granted.

Hazardous waste of 0.83 lakh MT produced per year by 478 units was disposed of by reprocessing (0.61 lakh MT), incineration (0.03 lakh MT) and on land (0.19 lakh MT). Details of action taken against 466 units working without authorisation, quantity of hazardous waste generated and its disposal were not available with the Board. The Board had neither prepared inventory of disposal sites nor assessed the environmental impact.

Municipal Solid Wastes (Management and Handling) Rules, 1999

The Municipalities were responsible for collection, segregation, storage, transportation, processing and disposal of Municipal Solid Wastes (waste). However, Municipalities did not process waste and bio-medical and hazardous waste was not segregated. Out of 3788 MT of waste generated during 1996-2001 in five Municipal Towns, 2829 MT was transported and dumped in dump areas outside the city limits and the balance quantities were left on road side etc. The dump areas were not protected with fencing/wall etc., except in Mangalore city. Thus, the areas around the dump were exposed to various health risks.

Board was to monitor compliance with Municipal Solid Wastes (Management and Handling) Rules (Rules) by Municipalities and grant them authorisation. The Board was also to obtain annual report from Municipalities. However, most of the Municipalities had not even applied for authorisation as of September 2001 and none of 219 Municipalities obtained authorisation.

These Municipalities generated waste of 2388 MT per day, but Board had no information about the manner of disposal of the waste. Only in June 2001, Board issued notices to all Municipalities for details. Thus, there was no monitoring of municipal waste by the Board.

Ineffective Monitoring of Vehicular Emission

Vehicular emission contains air pollutants like carbon monoxide (CO), Hydrocarbon (HC), Oxides of Nitrogen (NO X), lead, Oxides of Sulphur and Sulphur. Pollution load due to emission of various pollutants into air by vehicles for average distance ranging from 40 to 207 kms covered per day was as follows according to World Health Organisation norms.

Data indicated that emission from vehicles contributed to air pollution on large scale.

To prevent and control air pollution due to vehicular emission, State Government ordered (September 1991 and September 1995) that Commissioner for Transport, Director General and Inspector General of Police and Board should check and ensure that vehicular emission conformed to standards prescribed in Acts/Rules.

They were required to furnish compliance report to Secretary to Government, Department of Ecology and Environment. The RTOs were to monitor vehicular emissions by using equipment and also to check the vehicles for possession of Pollution Under

Control (PUC) certificate. As against the target of 16.55 lakh vehicles for the whole State prescribed for check by Commissioner for Transport for emission monitoring, 8.55 lakh vehicles (52 per cent) were checked during 1996-98 for possession of PUC certificates issued by authorised emission testing centres.

No targets were prescribed by the Commissioner for Transport from 1998-99 onwards and no compliance report was sent by him to Government. Consequently, percentage of vehicles checked for possession of PUC, declined significantly to 14, 12 and 8 per cent of registered vehicles during 1998-99, 1999-2000 and 2000-01 for the State as a whole.

Even while the level of pollutants in ambient air in Bangalore city increased one to four times (263 µg to 1892 µg per cubic metre) during 1998-2001, the number of vehicles checked for possession of PUC declined from 21 per cent (1998-99) to 13 per cent (2000-01) of the vehicles registered and in use. In 11 districts/regions less than one per cent of vehicles were checked while in other 12 districts/regions it ranged from 1 to 5 per cent of the vehicles registered during 2000-01. RTOs in 3 (Davanagere, Dharwad and Raichur) test-checked districts did not directly monitor vehicular emission.

The RTOs (Hassan, Raichur and Mysore) stated (May 2001) that emission could not be tested as equipment (smoke meter, gas analyser) used to go out of order frequently. Scrutiny revealed delay of 10 to 24 months in arranging repairs and RTOs failed to initiate timely action. Thus, monitoring vehicular pollution in Bangalore and other cities in the State was grossly inadequate and ineffective.

Other Points of Interest

(i) Non-recovery of Consent Fees

(a) Consent fee payable by industries etc., under the Acts/Rules was revised in December 1996 and reduced in April 1998. Arrears of consent fees due to revision for the State as a whole was not available with the Board. Scrutiny of records revealed that consent fee of Rs.27.20 lakh for the period from December 1996 to March 2001 in accordance with revised rate had not been paid by 53 units in test-checked districts. The regional officers stated that notices were issued for recovery of the dues.

(b) Failure of the Board to bring 2308 industrial units and 466 units generating hazardous waste within the purview of the Acts and Rules resulted in non-realisation of consent fee of Rs.37.26 lakh for one year.

(c) The rate of consent fee leviable on any industrial unit depended on capital investment. However, in respect of 169 software units located all over the State, consent fee was levied with reference to investment on diesel generators instead on total investments. It was stated by the Member Secretary of the Board that proposal had been sent (June 1999) to Government suggesting amendment to rules. However, the proposal had not been approved. The decision of the Board to levy consent fee with reference to investment on diesel generators only without approval from Government was contrary to rules. This resulted in loss of consent fee of Rs.5.13 lakh for one year.

(ii) Non-installation of Laboratory Equipment

Out of Rs.2.70 crore spent by the Board on purchase of various equipment under the programme of upgradation of laboratories implemented with World Bank assistance, 8 equipment costing Rs.55.06 lakh were not installed in three laboratories (Dharwad, Mysore and Davanagere) even after one to two years of purchase.

The Divisional officers in charge of laboratories (Davanagere) stated that firms who supplied equipment had not responded for installation inspite of repeated reminders and in one laboratory installation was pending for want of accommodation (Mysore).

(iii) Poor working of Laboratories

Board had established seven regional laboratories and one central laboratory in the State. The primary objective of these laboratories was to analyse water and stack samples and report results of analysis to respective ROs. As against 45853 (19372 during 1999-2000 and 26481 during 2000-01), water and stack samples required to be collected as per norms (February 1996), only 12877 samples were collected and analysed resulting in shortfall of 32976 samples (72 per cent).

The RO attributed (April to May 2001) shortfall in samples analysis to inadequate staff. The reply was not tenable as 70 per cent of sanctioned strength in scientific and technical cadre were in position.

(iv) Non-submission of Environmental Statements

Industries were to furnish to Board annual environmental statements which contained, inter alia, details of pollution and its control method. Out of total 6762 units (industries 6284 and hazardous waste generating units-478), only 529 submitted environmental statements to the Board.

The Member Secretary of the Board stated that there was no provision for initiation of penal action, but submissions of such statements was being insisted upon. In view of the poor response by the units, the necessity of provisions for initiating penal action needs to be considered.

Monitoring

Monitoring of implementation of Acts/Rules by Secretary to Government, Ecology and Environment was inadequate as:

(a) 2623 industries and 466 industries producing hazardous waste and 33 HCEs producing bio-medical waste had been operating without consent and authorisation respectively in violation of Acts/Rules,

(b) Stone crushing units were not covered under the Acts/ Rules,

(c) Proper functioning of AAQMS set up in Bangalore city could not be ensured,

(d) Instructions contained in Government order of September 1991, regarding inspection of vehicles for emission tests were not complied with. Management of bio-medical and hazardous waste was poor while compliance with standards for municipal solid waste management by Municipalities was not monitored at all. Pollution load due to emission from automobiles during 1996-97 to 2001 was not assessed though it significantly contributed to air pollution and Board was also required to monitor vehicular emission. Thus, Secretary, Board and other authorities did not ensure compliance with standards prescribed for various air and other pollutants.

3

KEY TECHNOLOGICAL ISSUES IN CONSTRUCTING AN ECOLOGY

The goal of constructing a well-off society on all fronts as put forward by the 16 Congress of the Party involves four targets, one of which is continuous enhancement of sustainability, improvement of the ecology and the environment, remarkable upgrading of resource efficiency, promotion of harmony between man and nature, and advancement of the entire society towards a civilized development path that features economic development, affluent livelihood and a sound ecology. As China faces enormous pressure of population growth and rapid development of the national economy, the construction of infrastructure is crucial to supporting socioeconomic development in the future. Grandiose projects such as the Three Gorges Project, the South-to-North Water Transfer Project, the West-to-East Pipeline Project and the Qinghai-Tibet Railway Projects are all of great significance to the realization of a well-off society on all fronts. The conflict between growing economic development and ecological/environmental protection becomes increasingly acute.

In a fairly long time to come, the development of water resources and hydropower will remain important infrastructure industries for the national economy. As such, water and hydropower projects capable of flood prevention, water supply, irrigation and power generation will play a significant supporting role in socioeconomic development. However, impacts of such projects on the ecological environment have attracted growing concern from the whole society, thereby becoming a constraint on such projects.

The Chinese energy mix has long focused on coal. In particular, coal-generated power takes up 70% of total power generation. The resulting huge emission of CO_2 and SO_2 not only severely pollutes the atmosphere but also subjects such power generation to constraint of international conventions. Water-energy is a clean and renewable energy. Aggregate water-energy resources in China top the whole world. After several decades' development, hydropower installed capacity has taken the first place in the world while hydropower generation is ranked No. 4 worldwide. Nevertheless, compared with developed countries, China is still limited in hydropower development, which is incompatible with the important status of water-energy resources in the entire Chinese energy mix. Consequently, China faces an arduous task in hydropower development.

China is rich in water-energy resources. The latest exploration data indicates that China possesses 689 million KW of water-energy reserves in theory (the 1980 census result was 680 million KW), 493 million KW (the 1980 census result was 378 million KW) installed capacity available for technical development, and 395 million KW installed capacity available for economic development. In order to satisfy the requirements of socioeconomic development in the new era, China should stick to the path of sustainable development, realize optimum allocation of resources within the largest possible extent, promote "west-to-east power transmission", highlight structural adjustment, protect the environment, save resources and promote healthy development of the power industry. All these need further acceleration of hydropower development and prioritizing the development of renewable water-energy resources.

However, the development of water-energy resources with common concepts and technologies has countered new constraints, i.e., ecological and environmental issues, which mainly include the following: sustainable development of the resettled population, impacts of constructing areas for the resettled on the ecological environment, the relationship with the natural reserves, impacts of reservoir sedimentation on rivers, impact of reservoir area silting on soil salinization, land sliding and earthquakes brought about by reservoirs, impacts of slope reclamation on vegetation and landscape, impacts of flood discharging scouring and atomization on bank slope, impacts of excavation spoils and concrete wastes on the environment, and impacts of dam inundation, blocking and run-off adjustment on biological resources, biological diversity and landscape diversification, especially on rare fishes. Hence, it is an imperative research task for China to coordinate water and hydropower development with protection of ecology and the environment along rivers or river basins.

The existing guiding principle of development is to consider ecological and environmental issues under the prerequisite of safety, technical feasibility and economic rationality. In order to realize the strategic objectives put forward by the 16 th Congress, China needs to attach more importance to ecological issues, and establish an environment-friendly hydropower project development system under which priority will be given to ecological and environmental issues throughout the stages of planning, exploration/testing, designing, construction, operation and management, so that China will reach a win-win situation of resource development/utilization and ecclogical/ environmental protection.

INTERNATIONAL TRENDS AND DEVELOPMENT

Developed countries have basically completed their infrastructure development. With an early start in hydropower generation, developed European and American countries reached 70% of water-energy resources development back in the 1950-60s. At the moment, developed countries basically refrain from large-scale development of the remaining 20-30% of water-energy resources. The situation in China, however, is different. At the moment, water-energy resources already developed and still under development only occupy 30% of the total quantity available for economic development. There is still 150-200 million KW of water-energy resources to be developed even if China is to reach the level of development that European and American countries achieved in the 1950-60s.

With growing awareness of environmental protection among the general public, some environmental issues resulting from the construction of hydropower projects have increasingly exhibited themselves. As such, how to take economic development and environmental protection into overall consideration has become a brand new task for developers of such projects. It is generally held that environmental issues possibly resulting from hydropower development include the following: After water impoundment of a reservoir, with the increase of industrial wastewater discharged into the reservoir, water flow of the reservoir area will slow down, diluting and diffusing capability of the water body will be reduced and concentration of pollutants in the water body will gradually increase, all of which will lead to decline of water quality in the reservoir; Dam construction will impact directly on water aquatic organism, especially on migrating fishes, and lead to the disappearance of spawning areas of some fishes; After impoundment, temperature mix of some reservoirs will change, which might cause cold damage on crops of the lower reaches.

In addition, after impoundment, changes in flow pattern will exert potential impacts on ecological environment of the lower dam and river outlet areas, and result in the rise of ground water level in periphery reservoir areas and salinization of soil. Since the 1970s, mankind has paid more attention to ecological and environmental issues caused by dam construction. Some scholars hold that the development value of clean, efficient, reliable and renewable hydropower resources is being reduced by its long-term and extensive potential negative impacts on ecological equilibrium. Inundation, blocking and change of run-off courses resulting from dam construction lead to fragmentation of the river ecological system, which will impact severely on the aquatic ecological system. Main expressions thereof include the following:

Loss of the original natural run-off and hydrological conditions that the original species have been used to;

Loss of functions of rivers as the communication corridor for organism and nutrients;

Decline of the "filtering" function of the connected high land and water-area ecological systems along the rivers. The extent of impact depends on the scale and position of projects. Some large-scale water and hydropower projects may affect the ecological formation of the entire river basin, leading to erosion and marine transgression at the lower-reach riverbanks and source reduction of nutrients in the outlet regions.

Research over the recent years focuses on the inclusion of ecological and environmental impacts into the whole process of project planning, designing and construction so as to mitigate the impact of water projects on the ecology and the environment. The US Department of Energy has developed the ecological turbine and established biological passes for protection of fishes.

RESEARCH LEVEL AND TECHNICAL STATUS QUO IN CHINA

Since implementation of the environmental assessment system, China has carried out considerable research on the impacts of large-scale project construction on the ecology and the environment, focusing on the Three Gorges and South-to-North Water Transfer Projects. Such research, however, is mainly confined to macroscopic qualitative statement rather than microscopic quantitative work, and is focused more on analysis of impact on the existing ecology and environment than on forecasting in terms of analyzing ecological and environmental effectiveness. For instance, at the moment, assessment of ecological and environmental effectiveness of large water projects in China is restricted to single-factor assessment, with very limited consideration given to functional changes of the ecology in the river basin after project implementation, structural and functional degradation of the ecological system in the basin and the corresponding recovery measures.

At the same time, environmental assessment of strategies and plans and the research of the service functions of the ecological system are at the initial stage while research on recovery of damaged ecological system is still at the qualitative stage. In face with the fragile ecology and environment and rapid economic growth, how to coordinate the conflicts between rapid economic development and ecological/environmental protection is an important aspect of sustainable development. The current consensus is that emphasis should be placed on ecological and environmental protection throughout the process of policy-making, planning and construction.

In addition, China is deficient in systematic research and consideration of the impact of dam construction on the survival and reproduction of fishes. With most rivers failing to consider the installation of passage facilities for fishes in dam construction, such deficiency is even more acute in the case of engineering measures. In particular, insufficient attention is paid to ecological and environmental protection and recovery, while more emphasis has been placed on returns on power generation, flood prevention and irrigation. Areas that require sufficient attention and substantial research include how to re-examine the objectives of developing water resources and hydropower in China from the perspective of sustainable development, how to give genuine consideration to intensive and practical attention to environmental and ecological protection and positive/effective engineering measures, and how to conduct salvage project research of issues that really impact the ecology and the environment.

RECOMMENDATIONS FOR TECHNOLOGICAL DEVELOPMENT

In order to establish a large-scale environment-friendly system for hydropower project development, China needs to focus on the development of the following research areas.

Review and Evaluation of Environment Assessment of Built Dams

Implications

It is of great significance to review and evaluate environment assessment of built dams, summarize the extent and degree of impact that project construction and operation exert on the ecology and the environment, and examine the effectiveness of measures that are adopted in project development for soothing out ecological impacts.

Such researches are conducive to systematic understanding of multiple effects on habitat, biological diversity, water quality and society that result from inundation, blocking and run-off changes due to dam construction and understanding of the "real' impacts of dam construction on the ecology and the environment. They are also of great significance for the production of pertinent measures for mitigating environmental impacts, and to the provision of systematic preliminary data for designing and constructing ecology-friendly dams in the future.

It is expected that the above research efforts will lead to the following results:

Objective, fair and scientific assessment of the advantages and disadvantages that the dams will bring to the ecology and the environment after several years' operation; Production of recommendations and measures for improving the ecology and the environment, enhancing dam management and operation and reducing negative impacts;

Summary of past experiences and lessons and establishment of future-oriented ideology, understandings and ways of thinking in pace with increasing economic, social and ecological requirements regarding project development such as dam construction.

Major Research Contents

Impact of dam construction on the ecology and the environment will be analyzed from planning and project development stages. Research work will be conducted on three levels, i.e., Carry out systematic survey of dam construction and its impact on the ecology and the environment in China;

Conduct analysis of sensitive issues in terms of the impact of dam construction on the ecology and the environment via case studies;

Produce ecological standards for ecology-friendly dam construction and operation.

Specific research contents are described below.

Establish a case base for China's dam construction and its impacts on the ecology and the environment. Efforts shall be made, under systematic survey, to select representative dams for overall consideration of elements such as dam size, development objectives, operation modes and ecological and environmental features of the rivers, analyze the impact of parameter portfolio on ecological functions and structure of rivers, so that a case base for dam construction and its impacts on the ecology and the environment will be set up.

Review and evaluation of dam construction and its impacts on the ecology and the environment Typical cases shall be selected to systematically analyze dam construction and its impacts on the ecology and the environment. In view of existing topical issues, efforts shall be made to conduct priority analysis and review of plans for cascade development of hydropower stations from the layers of planning and project development, and positive/negative impacts of reservoir (power station) construction on the ecological environment.

Systematic research shall also be carried out in terms of cumulative impacts of typical river-basin dam construction on the river-basin ecology and environment, while evaluation shall be made of the effectiveness of measures already adopted for mitigation of ecological impacts. Review and analysis study shall also be done for rationality of river-basin development plans in terms of their impact on the ecology and the environment.

Review and evaluation of changes in hydrological patterns resulting from dam construction and their impacts on the habitat and the environment Through comparison of hydrological data before and after dam construction, summary shall be made of changes taken place in important ecological and hydrological indicators. In combination with species changes of indicative organism, efforts shall be made to analyze impacts of habitat changes on the growth history, species and quantity of organism.

Alternatively, comparison of remote data series will be used to analyze habitat changes. Research of ecological standards for ecology-friendly dam construction On the basis of review and evaluation and in combination with domestic and overseas research achievements, basic ecological standards shall be established for dam construction and reservoir operation, including theories and methods for determining minimum quantity of undershot ecological flow quantity, ecological and hydrological scheduling for ecological recovery of river adaptability, and new-style reservoir scheduling standards that are based on the integration of ecological, hydrological and project scheduling.

Research of Techniques related to Environmental Assessment of Plans for Developing Water Resources and Hydropower

Implications

It is required by both governance and river-basin development and utilization to carry out environment assessment of water resources and hydropower planning. Such assessment is basically a blank space in China. With only some river basins or regions carrying out conceptual work, China lacks a complete system of assessment theories, techniques, methodologies and standards. At the same time, due to the increasing prominence of ecological and environmental problems in different regions over the recent years, environmental assessment of regions, plans and strategies have remained topical issues of environmental science and researches.

We shall also carry out analysis and assessment of environmental cost for development of water resources by observing the principle of sustainable development of river-basin environment, and in accordance with the natural conditions, functionality, multiple objectives, dynamics and uncertainty of the river-basin environment. Dynamic development modes should be used to analyze economic and environmental elements while environmental systematic planning and integrated management will be used as the main indicators for gradual establishment of a set of theories and methodologies for river-basin planning assessment that combines qualitative with quantitative elements, so as to improve and innovate dynamic development modes that are geared to sustainable development of the environment and set up an indicator system for environmental assessment of planning.

Active preliminary work in this regard shall be of great significance for providing water authorities with operable management basis for the latter's implementation of the Environmental Assessment Law, and for formulating water and hydropower plans and carrying out river-basin ecological and environmental protection in the new era.

Major Research Contents

Scope of Environmental Assessment Concerning Water Resources and Hydropower Planning and Analysis and Research of Key Ecological and Environmental Issues

For determination of the scope and priorities of environmental assessment concerning water resources and hydropower planning, it is necessary to classify such planning activities, analyze their direct and indirect impacts on the ecology and the environment, investigate and analyze environmental problems caused by different planning, and analyze and study key ecological and environmental issues related to special water resources and hydropower planning under uniform total river-basin management of water quality and quantity.

Research of Theories and Methodologies Concerning Ecological and Environmental Carrying Capacity of River Basins (or regions)

As the prerequisite for rational planning of water resources and hydropower, the assessment and study of ecological and environmental carrying capacity include research of the establishment of an assessment indicator system and subsequent research of theories and methodologies concerning ecological and environmental carrying capacity.

Cumulative Effect of River-basin Cascade Development and Techniques/methodologies for Assessing such Effect

Relevant work in this regard includes systematic study of cumulative impact of dam construction on river-basin ecology, environment, natural environment and social/economic aggregates, as well as the exploration of key elements related to environmental impacts caused by river-basin development. One single plan might only exert very limited impact on the ecology and the environment. However, accumulation of different development activities usually results in summed effect or synthesis.

Assessment of cumulative effect is an issue that must be considered at the stage of planning for river-basin development, which involves the study of techniques and methodologies for assessing such cumulative effect of river-basin development and utilization and the subsequent production of such techniques and methodologies.

Establishment of a System of Criteria for Environmental Assessment of Water Resources and Hydropower Planning

The establishment of a system of criteria for environmental assessment of water resources and hydropower planning includes the collection of information related to techniques and criteria of such assessment, domestic and overseas comparative study of such techniques and criteria, as well as the establishment and improvement of a system of criteria for environmental assessment of water resources and hydropower planning.

Research of the Integration of Techniques Related to Environmental Assessment of Water Resources and Hydropower Planning

With regard to research of the integration of key techniques related to environmental assessment of water resources and hydropower planning, it is necessary to design a mode that integrates the techniques and methodologies with the system of technical criteria for environmental assessment of water resources and hydropower planning, integrate and establish river-basin spatial information data management technologies and the methodology base for such environmental assessment, and carry out research of key technologies.

Ecological and Environmental Effects of Water and Hydropower Projects and the Countermeasures Thereof

Implications

In a fairly long time to come, development of water resources and hydropower will remain important infrastructure industries for the national economy. As such, water and hydropower projects capable of flood prevention, water supply, irrigation and power generation will play a significant supporting role for socioeconomic development. However, impacts of such projects on the ecology and the environment have attracted growing concern from the whole society, thereby becoming a constraint on such projects. Elements that water and hydropower projects cannot afford to ignore include impacts of dam inundation, blocking and run-off adjustment on biological resources, biological diversity and landscape diversification, the risk to biological safety as a result of inter-river-basin water transfer and impact on the lower-stream biology. Hence, it is an imperative research task for China to coordinate water and hydropower development with protection of the ecology and the environment along rivers or river basins. With the issue of project technologies already taking its preliminary shape (including the introduction of overseas technologies), ecological and environmental issues are major constraints on future project development.

Content of Research

Research of Mechanism Concerning the Impacts of Changes in Key Hydrological Elements on Biological Resources

From the macroscopic perspective, comparison will be made of long-term and extensive changes in key hydrological elements and the rules of death and growth of biological resources, so as to study the extent and ways of hydrological patterns that are caused by water and hydropower projects and their impacts on biological resources. From the microscopic perspective, efforts shall be made on the basis of different organism's attraction to and avoidance of different hydraulic conditions, so as to study the impact of different hydraulic environments (such as velocity, flow pattern and run-off adjustment below dam) formed as a result of the related water and hydropower projects on important biological resources and explore the functions of such projects and the biological and hydraulic mechanism of important biological resources.

Research of Major Compensations for Negative Impacts on Important Biological Resources

Efforts should be focused on the mainly impacted species and the key sources of negative impacts, i.e., Technologies for rebuilding the habitat for key biological resources (man-made habitats); Technologies for adjusting and compensating run-off on the basis of biological hydraulics (such as projects for man-made flood peak and undershot temperature adjustment and technologies for run-off adjustment); Technological development of eco-hydraulic engineering.

Research of Biological Standards Concerning the Operation of Ecology-friendly Water and Hydropower Projects (including bio-hydrology and bio-hydraulics)

In this regard, efforts shall be made to conduct further research and establish basic biological standards for dam construction and reservoir (power station) operation, including theories and methods for determining minimum undershot biological flow quantity, bio-hydrological scheduling for biological recovery of river adaptability, and new-style reservoir scheduling standards that integrate bio-hydraulics with project scheduling.

Exploration and Research of Ways for Eco-friendly Investment-oriented Resettlement

Efforts shall be made to take advantage of the opportunity of water and hydropower development and further explore new-style resettlement policies for sustainable development.

At the same time, modes of eco-friendly investment-oriented resettlement shall be studied according to the indicator and principle that the ecology and the environment will be improved while the number of resettled population and area of inundated farmland per unit installed capacity or reservoir capacity will be reduced.

Research of New Technologies for Project Exploration and Testing based on 3S Technologies and Physical Exploration Technologies

Implications

At the moment, geological explorations such as tunnel, dent and pit explorations are frequently applied in exploration for water and hydropower projects, which, due to explosion, excavation, random placement of spoils and omission of backfill and sealing, has to some extent impacted on the ecology and the environment of the project area. As such, we should vigorously develop project exploration technologies that are of less impact on the ecology and the environment, such as the 3S (RS?GPS?GIS), as well as new physical exploration technologies that are represented by integrated well-testing technology, micro-earthquake technology, shallow reflection, shallow refraction, chromatography imaging technology, reflective imaging, magnetic-wave geological radar technology, VLF magnetic wave testing technology, etc.

Key Research Content

(1) Integrated technical research on the application of 3S system in exploration and testing of water and hydropower projects;

(2) Research into ways and methods for establishing GIS visualized 3D spatial data model for geological information in project exploration areas;

(3) Research on technologies for interchange of 2D and 3D geological information;

(4) Research on analytic techniques concerning 3D geological information models for specific geological bodies;

(5) Research of technologies for integrated utilization of 3S exploration and testing results (such as quality assessment of rock mass, analysis of slope stability and automatic imaging system of development project exploration and testing);

(6) Research of the adaptability, reliability and accuracy of various physical exploration technologies

Research of New Technologies for Careful Design, Excavation and Consolidation

Implications

Since the 1970s, with the construction of a considerable group of large and medium-sized hydropower stations, excessive tunnel, dent and pit excavation and slope reclamation have overwhelmingly changed the original natural topography and land form, destroying and negatively affecting the ecology and the environment, which is incompatible with the requirements for building large ecology and environment-friendly power station projects.

In order to minimize the impact of project development on the natural environment, it is imperative to carry out research of new technologies for careful design, construction and consolidation of projects.

Key Research Contents

(1) Research of key technologies for large-scale underground plants, including:Research of advance forecasting technologies concerning excavating geology of tunnels during the construction stage; research of new technologies for excavation of large underground tunnels under complicated geological conditions; research of new technologies for lining and support and tunnel formation of large underground tunnels under complicated geological conditions; research of new technologies for non-damage quality testing of lining and support construction for large underground tunnels; research of new technologies for feedback information and decision

making concerning stability monitoring of adjoining rock of large underground tunnels; research of new technologies for excavation of large underground tunnels; research into methods for analysis and integrated assessment of the stability of adjoining rock of large underground tunnels under complicated geological conditions;

(2) Fine technologies for excavation of slopes, including: research of technologies for integrated control of high slopes; research of technologies for ecological restoration of high slopes as a result of certain projects; research of technologies for total utilization of materials used in high slope excavation; research of new technologies for slope exploration and monitoring; integrated analysis of project slope stability.

(3) Research of explosion and excavation technologies under complicated environmental conditions, including: research of methodologies for controlled explosion of outline excavation under complicated environmental conditions; development of imitation systems aimed at underground project excavation and explosion; research of technologies of fine excavation and explosion for tunnel formation and outlet locking; research of technologies for total-utilization controlled explosion that integrates with explosion-resistant research and excavation materials that meet requirements of dam fillers; research of criteria for assessment and control of safety impact of explosions; analysis of technologies for controlled explosion in terms of their impact on high slope stability and dynamic stability.

Research of Key Technologies for Preventing and Controlling Geological Disasters Caused by Reservoirs

Implications

Reservoir impoundment constitutes certain destruction on the stability of the banks, which is mainly expressed in the rise and fall of reservoir water level and the impact of wave scouring on the banks. At the same time, reservoirs result in the rise of ground water along the banks and cause changes in the nature of stress and physical dynamics within the bank rock mass and cliff debris. Softening of rock soil and decline of intensity will lead to a series of geological disasters along slopes, such as slide of bank cliff debris, revival of ancient landslips and collapse of bank slopes. Such geological disasters have taken place to different extents in built water and hydropower projects, exerting certain impacts on the ecology and the environment of the reservoir areas. China began its forecasting and study of landslides and avalanche in the 1960s, and has since put forward a variety of slide forecasting methods including but not limited to the three categories of determinative forecasting model, non-determinative forecasting model, and comparative analysis.

However, it remains unresolved in China as to what theories and theoretical models should be adopted in forecasting.

China is still in the process of exploring for effective solution of and theoretical models for forecasting landslides and avalanches.

Key Research Content

(1) Preliminary theoretical research related to forecasting of landslide and avalanche disasters in reservoir areas;

(2) Automatic systems for early warning and 3D monitoring of GPS-based landslides (including regular monitoring, laser positioning 3D monitoring, photographic imaging analysis and monitoring, earthquake chromatography imaging 3D monitoring); Efforts shall be focused on practical GPS technologies; forecasting software for real-time receipt, disposal and analysis of monitoring data and data bases for such monitoring.

(3) Research of methods for analyzing landslides and slope stability;

(4) Research of optimum measures for integrated control of reservoir collapses and landslides

Utilization of Abandoned Excavation Materials, Filling and Building Criteria and Technologies for Reclaiming Construction Sites

Implications

In traditional hydropower development, large amount of earth-rock projects inevitably explode mountains and rocks with earth taken for filling and building purposes, which destroys the surrounding ecology and environment to some extent. At the moment, such extensive practice will undoubtedly be subject to the constraint of environmental protection. An urgent task for the new generation of engineers is how to build ecology-friendly hydropower projects.

With regard to the previous research results, China has done a series of studies concerning rational utilization of earth-rock materials in its hydropower development. The Chinese starting point, however, has been based on economic considerations or failed to consider environmental elements in a conscious manner due to the shortage of sources of materials. Consequently, the Chinese research results are not comprehensive. Nevertheless, relevant research results have also laid a good foundation for further study in the future.

In face with huge area covered by construction sites of hydropower projects, China should pay attention to ways of reclaiming and utilizing the sites after project completion.

Key Research Content

(1) Research of the engineering mechanics of poor earth-rock materials, the feasibility of applying such materials in filling dam bodies and technologies for such dam building;

(2) Research of the impacts of different dam materials distribution areas on deformation of dam body stress, as well as the reasonable shape of dam structure when building excavation materials or other poor earth-rock materials are used to fill up the dam body;

(3) Design scheme as well as technical criteria for construction control in replacing clay with poor earth-rock materials to fill up the core wall of an earth-rock dam;

(4) Technologies for ecological restoration of material premises and reclamation/utilization of construction sites

Research of New Technologies of Energy Dissipation for Mitigation of Scouring Caused by Flood Discharging, Rain and Fog

Implications

In flood discharging, high dam projects usually apply surface holes, medium (deep) holes, spillways and spillway tunnels. For the purpose of enhancing efficiency of energy dissipation, nappe collusion of holes and medium (deep) holes is also adopted to dissipate energy. However, scouring caused by flood discharging, rain and fog remains a serious problem. From the perspective of protecting banks and meeting specific environmental requirements, it is necessary to research into new energy-dissipating technologies that are efficient in energy dissipation, minor in atomization impacts and economically viable.

Key Research Content

(1) Mechanism research of flood discharging rain and fog;

(2) Energy-dissipating technologies within flood-discharging holes;

(3) New energy-dissipating technologies that integrate the construction of flow-diverting holes with permanent flood-discharging holes

Technological Research of High-dam Fish Passes or Fish Stairs

Implications

Amongst other impacts on the ecology and the environment, dam construction affects growth and reproduction of fishes, which was not taken into sufficient consideration in previous dam construction in China. While developing the Gezhou Ba Hydraulic Project, a huge amount of research was done to protect the Chinese sturgeons.

At present, China has made some achievements by way of artificial reproduction of the sturgeons. In the overall sense, we have not paid sufficient attention to fish protection in dam construction, let alone the conduct of corresponding researches. As such, further efforts should be exerted in this regard.

Major Research Content

(1) Efforts should be made to survey the existing living situation of fishes in typical rivers (rivers with stair development of dams or basically without development), including investigation and analysis of fish distribution prior to development from the perspective of geographic distribution conditions, with focus placed upon changes of the water body environments at upper and lower reaches due to reservoir construction, elements that affect survival and reproduction of fishes and the degree of such effect and survey of hydraulic conditions required by different fishes in different localities for survival and reproduction.

(2) Survey and study the design, utilization, management and technical standardization of domestic and overseas buildings that help fishes pass dams (especially regarding high-dam construction), focusing on overseas stair-development projects along rivers. On the basis of overseas research and application experiences and lesson, efforts shall be made to define problems that exist in protecting dam-passing fishes in China's dam construction and put forward the direction for future research.

(3) Carry out research of hydraulic conditions of buildings that are geared to the passage of fishes through dams (overflow turbines), including physical model experiment and numerical imitation research from the perspectives

of streamlining structures, improving convenience of construction and management.

(4) Develop project measures for addressing living conditions of fishes in reservoir areas and impacts of dissolved oxygen from lower stream buildings on fishes.

(5) Select typical completed projects (at the moment medium or small projects would be better choices) to carry out applied research of trial projects and follow-up observation, including the input of fries, the growth, migration and reproduction of fishes, with elements considered covering water temperature, velocity, sedimentation content and physical/chemical parameters.

ECOLOGY AND ENVIRONMENT OF MANGROVE ECOSYSTEMS

The mangrove environment has some special physicochemical characteristics of salinity, tidal currents, winds, high temperatures, and muddy anaerobic soil.

Plants of the environment are able to adapt to themselves to practically all types of adverse conditions except perhaps frost, and hence they are distributed mostly in the tropical regions. Probably there are no other groups of plants with such highly developed adaptations to extreme conditions. Mangroves occur in low-lying, broad coastal plains where the topographic gradients are small and the tidal amplitude large. Repeatedly getting flooded, but well-drained soil supports a rich growth of mangrove plants.

They normally grow poorly in stagnant waters and have luxuriant growth in the alluvial soil substrates with fine-textured loose mud or silt, rich in humus and sulphides. They can also be found in substrates other than muddy soil such as coastal reefs and oceanic islands. In such areas, the mangrove plants grow on peat, which is derived from decayed vegetation. They find it difficult to colonize the coastal zone with waves of high energy and hence they normally establish themselves in sheltered shorelines (Kathiresan & Bingham, 2001).

TYPES OF COASTAL SETTINGS

Mangroves get tightly bound to the coastal environments in which they occur. Not only are they influenced by chemical and physical conditions of their environment, but they usually help to create those conditions by themselves. They are found in a variety of tropical coastal settings like the deltas, estuarine areas with their own deltas, lagoons, and fringes of the coral reefs.

There are normally six functional types of mangrove forests, namely, fringe, riverine, basin, overwash, scrub (dwarf) and hommock forests. The last three types are the modified forms of the first three types. The six types can be summarized as follows:

1. *Overwash Mangrove Forests:* These are small mangrove islands, frequently formed by tidal washings.
2. *Fringing Mangrove Forests:* These occur along the borders of protected shorelines and islands, influenced by daily tidal range. They are sensitive to erosion and long exposure to purely marine conditions with turbulent waves, and tides.
3. *Riverine Mangrove Forests:* These are luxuriant patches of mangroves existing along rivers and creeks, which get flooded daily by the tides. Such forests are influenced with the incursion of large amount of freshwater with fluvial nutrients and thus making the system highly productive with trees growing taller.
4. *Basin Mangrove Forests:* These are stunted mangroves located along the interior side of the swamps and in drainage depressions. Their positioning channels the terrestrial runoff to move towards the coast with a slow velocity of water flow.
5. *Hammock Mangrove Forests:* These are similar to the basin type except that they occur in more elevated sites than the four types given above.
6. *Scrub Mangrove Forests:* These form dwarf mangrove settings along flat coastal fringes.

Based on the substrate, tidal range and sedimentation, six more broad classes of mangrove settings have been given as follows by Thom (1982) and Galloway (1982):

1. Large deltaic systems (occurring in low tidal range, very fine allochtonous sediments) (e.g. mangroves of Borneo, Sundarbans)
2. Tidal plains (where alluvial sediments are reworked by the tides; and there is the presence of large mudflats for the growth of mangroves)
3. Composite plains; under the influence of both tidal and alluvial conditions (e.g. lagoons formed behind wave-built barriers where mangroves grow)
4. Fringing barriers with lagoons (high wave energy conditions with autochtonous sediments of fine sand and mud) (e.g. mangroves of the Philippines),
5. Drowned bedrock valleys (e.g. mangroves of Northern Vietnam or Eastern Malaysia)
6. Coral coasts (mangroves growing at the bottom of coral sand or in platform reefs) (e.g. mangroves of India, Indonesia and Singapore)

Most of the types of mangrove forests noted above get frequently affected by rapid changes in coastal geomorphology. Many ecological factors strongly influence the well-being of mangroves and these include geographical latitudes, wave action, rainfall, freshwater runoff, erosion/sedimentation rates, aridity, salinity, nutrient inputs, and soil quality. Primary productivity and biomass of mangroves normally decrease with increasing latitudes.

ECOLOGICAL CHARACTERISTICS TIDAL GRADIENT

In general, coasts with a greater tidal range can help in colonization of extensive mangrove forests. A greater tidal range increases the intertidal area and depending on the slope of the substrate, which if smooth encourages the growth of mangroves. The topography of a mangrove swamp is normally smooth and flat. However, any rise or fall of the land can greatly influence the direction and rate of flow of the water in the system, which changes the colonization of mangroves.

Many species of mangroves respond differently to different tidal regimes. For example, in the Sundarban area of India, a mangrove forest that experiences total diurnal inundation is dominated by Avicennia marina and A. alba while Excoecaria agallocha, Ceriops dacandra and Acanthus ilicifolius get dominated at sites that are not completely inundated. Nypa fruticans also seems to prefer sites with low level of tidal inundation. If diversity is high, mangroves get extremely variable in their location. Their "zones" may be obscured by a broad overlap in species distribution or they may simply remain absent in some mangals.

Temperature

Mangroves attain climax growth only under tropical conditions where atmospheric temperature in the coldest months is greater than 20°C and the seasonal fluctuation does not exceed 5°C. Mangroves have been reported to grow in latitudes where the average sea surface temperature is 24°C. Any further rise in temperature may lead to spreading of only some species in to higher latitudes, provided that the direction of ocean currents facilitates the dispersal of their seeds. However, very high temperatures are not favourable as leaves of mangroves are sensitive to temperature and their photosynthetic capacity gets reduced, falling to zero at leaf temperatures of 38-40°C, as against the optimum leaf temperature for photosynthesis which is 28-32°C. A small increase in temperature may not adversely affect the flowering, but may change their reproductive cycle and thus may alter the duration between flowering and the fall of ripe seeds.

Global warming may lead to sea level rise and as a consequence of that, mangroves may shift more landward. However, in many parts of the world, mangroves are unlikely to migrate landward because of human settlement at the landward boundary and the width of the mangrove forests is likely to decrease under the rise in sea level. The effects of sea level rise on any mangrove habitat will be influenced by local conditions such as the wetland type, geomorphic setting, and human activities in the wetland. It is predicted that the sea level rise may range from 45 to 65 cm /100 year. However, mangroves could survive a rise rate of 8-9 cm/100 year and hence the predicted faster rates could seriously threaten mangrove ecosystems. This view was challenged by Snedaker et al. (1994) who cited historical records showing mangrove expansion under relative sea level changes nearly twice as great. Tan and Zhang (1997) conclude that, given the predicted rates of estimate, the sea level rise poses no serious threat to most mangrove forests in China.

Salinity

Salinity plays a vital role in the distribution of species, their productivity and growth of mangrove forests. Changes in salinity are normally controlled by climate, hydrology, rainfall, topography and tidal flooding.

Mangroves generally tolerate higher salinity than do non-mangrove plants, but tolerance also varies among the mangroves. For example, Rhizophora mucronata seedlings do better in salinities of 30%, but R. apiculata do better at 15%. Sonneratia alba grows in waters between 2% and 18%, but S. lanceolata only tolerates salinities up to 2%.

In general, mangrove vegetation is more luxuriant in lower salinities. Experimental evidences indicate that at high salinity, mangroves spend more energy to maintain water balance and ion concentration rather than for primary production and growth. On the Pacific coast of Central America, freshwater availability (largely from rainfall and land runoff) controls reproductive phenology, growth and mortality of Avicennia bicolor.

However, low salinity, associated with long periods of flooding, contributes to mangrove degradation through reduced cell turgidity and decreased respiration. Mangroves are poor competitors under non-saline areas where freshwater marsh plants easily out-class them. To cite an example, in Americas, the riverine mangrove forests disappeared at the Amazon and

Orinoco river estuaries, where freshwater flow dominated over the tidal prism and under this situation; freshwater macrophytes replaced mangroves.

Salinity at high levels also affects mangroves. For instance, high salinity reduces the biomass in hydroponically grown Bruguiera gymnorrhiza, and causes denaturing of terminal buds in Rhizophora mangle seedlings. Saline interstitial water is known to reduce the leaf area, increases the osmotic pressure leaf sap, increases the leaf area/weight ratio, and decreases the total N, K, and P minerals. Simple salinity fluctuations also have significant negative effects on the photosynthesis and growth of plants. In Senegal, hypersalinity (from a decade of low rainfall and high evaporation) has caused the growth of salt marshes invading into mangrove areas and thus completely destroying their vegetation. Extremely high salt concentrations in the groundwater of tropical salt flats are also responsible for the complete absence of mangroves in such locations. There are often very sharp changes in the groundwater salt concentrations at the interface between salt flats and mangroves indicating that the mangroves are modifying the salinity of the groundwater.

Mathematical models developed for the groundwater flow show that human activities at hundreds of kilometers inland can destroy vast areas of mangroves by changing the flow of groundwater and thus modifying the salinity levels.

RAINFALL AND SUPPLY OF FRESHWATER

The availability of freshwater is an important factor for the development and growth of mangrove forests. Freshwater supply has often been indicated by the ratio of rainfall to evapotranspiration. Under the humid conditions, where the ratio exceeds 1, the mangroves grow luxuriantly. In arid climates, on the other hand, where the ratio falls below 1, the mangroves get stunted. High rainfall in humid areas leaches out residual salts from the mangrove soil and thus encourages the growth of mangroves. For example, along the Panama coast and southern coast of Costa Rica, where the annual rainfall ranges between 2100 and 6400 mm, mangroves exceed 35 m in height and have a biomass of 280 t ha-1. In the arid regions, barren salt flats often develop along the landward rim of the mangroves, due to poor leaching of salts from the soil due to low rainfall and this result in poor growth of mangroves because of high salinity of the soil.

Sediment Characteristics

Three conditions favour mangrove formations and these are:

1. Open seashores where substratum is muddy and also protected,
2. Protected banks of estuaries, and
3. Creeks which cut across the land mass. Among these three, protected banks of estuaries are the best suited locations for the mangroves.

One of the important factors in the mangrove habitat appears to be nutrient concentrations. Nutrient fluxes in these environments are closely dependent upon plant assimilation and microbial mineralisation. Two major elements (N and P) are of great significance for the growth of mangroves. The concentration of dissolved inorganic nitrogen is generally low in tropical mangrove waters. Some microorganisms incorporate N_2 from the atmosphere and convert the nitrogen present in the soil into ammonium, which makes nitrogen available for the use of plants. Ammonia is the main form of inorganic nitrogen in mangrove soil. It combines with oxygen to form nitrite or nitrate by the process of nitrification, which is used by the plants. This process normally occurs in the root zone that releases oxygen. Like nitrogen, the concentration of inorganic phosphorus is also generally low in mangrove waters. It is normally available as a soluble salt that can readily be assimilated by the plants. Phosphate is efficiently adsorbed by the fine sediments of muddy areas rather than the coarse-grained sediments. This is probably the reason for mangroves growing luxuriantly in muddy environments.

Nutrient availability may limit growth and production in many mangals. Pulses of high nutrient can create immediate and significant changes in the vegetation. For example, on the southeast coast of India, seedlings grow 5 times as high and produce 4 times as many leaves in the post-monsoon season as they do in the dry season. The reason is attributed to high nutrient concentrations and moderately low salinity during the late monsoon period.

The limitation in nutrients and trace metals varies with individual mangrove habitats. For example, potassium levels may be important in some regions. Rhizophora apiculata seedlings do significantly better in plantation sites enriched with potassium. In general, the growth of mangroves in carbonate soils with low-nutrients gets limited in phosphorus-impoverished situations. This happens because whatever phosphorus may be present gets bound with calcium, and is effectively held within the sediments. In mesocosm and field experiments with Rhizophora mangle seedlings, phosphorus enrichment produced a nearly 7-fold increase in stem elongation rates and a 3-fold increase in the leaf area. Nitrogen addition produced no such response. Similarly, low phosphorus availability limits the growth of dwarf R. mangle in a Belizean mangal and promotes the develorment of hard, long-lived leaves called sclerophylls.

The sclerophylls are probably just an adaptation to conserve the nutrients in these oligotrophic habitats. Mangroves may develop other types of mechanisms to conserve valuable nutrients. For example, mature photosynthetically active leaves have much higher nitrogen and potassium concentrations than senescent leaves. This is apparently a result of nutrients being translocated from the ageing leaves and into other parts of the plant before the leaves fall. Damage to mangroves declines the levels of nutrients. For example, severely damaged mangrove sites in North Queensland led to a significant loss of both nitrogen and phosphorus from the soil. This may be indirectly related to decline in crab population as indicated by a dramatic decrease in the density of burrows. The effect of disturbance differs from habitat to habitat and largely depends on the characteristics of sediments and the flow regime of nutrients in each site.

For example, Triwilaida & Intari (1990) found that there are significant differences in the nutrient concentration of soil between the degraded and healthy mangrove stands. Kathiresan (2002) recorded low levels of available nutrients, high salinity and low microbial load in degraded mangrove soil, as compared to those in the soil of luxuriant mangroves. Mangrove soils are in general acidic in nature; especially in those areas which get flooded infrequently and are located in the upper areas of the intertidal zone. Soil becomes loose because of the presence of fine sediment and decaying organic matter. In such conditions, the soil becomes black in colour and emits foul smell of H 2 S due to anaerobic microbial activity in the soil.

The occurrence of sulphides is a characteristic feature of mangrove sediments and it influences the distribution of mangroves. For example, in a neotropical Florida mangal, the zone dominated by Rhizophora mangle (with its numerous aerial prop roots) has moderately reducing soil with low sulphide levels. In contrast, the zone of Avicennia germinans has strongly reducing soils with high sulphides. In general, high sulphide levels can damage, reduce growth and cause high mortality of mangroves. Sulphide is formed due to reduction of sulphate and it is generally slower in young forests as compared to the older forests because of the occurrence of high nutrient level and low sulphide toxicity. Disturbance can increase the rate of sulphide reduction. Clearing of mangrove forests or simple formation of canopy gaps can change the physical and chemical characteristics of the underlying soil, leading to anaerobiosis and an increase in the sulphide activity in the sediment. Heavy organic input can also increase sulphide production and the H_2S accumulation usually can kill the mangroves if their pneumatophores get covered by silt and are unable to transport oxygen to the rhizospheres. Because of the presence of aerial roots, Rhizophora species have better chances of survival even in aged mangrove soil high in H 2 S. They thus form natural successful competitors to the less-tolerant Avicennia species.

4

THE SIGNIFICANCE OF CONTEXT IN COMMUNITY-BASED RESEARCH

Community workshops are widely used tools for collaborative research on social-ecological resilience in indigenous communities. Although results have been reported in many publications, few have reflected explicitly on the workshop itself, and specifically on understanding what is said during a workshop. Drawing on experience from workshops held in Huslia, Alaska in 2004 on wildfire and climate change, we discuss the importance of considering cultural, political, and epistemological context when analyzing statements made by indigenous people in community workshops. We provide examples of statements whose meaning and intent were, and may remain, unclear, with descriptions of our attempts to understand what was being said by placing the statements in a variety of contexts. We conclude that, although workshops can be an efficient means of exchanging information, researchers should strive for multiple channels of communication and should be cautious in their interpretations of what is said.

INTRODUCTION

Working with local residents has become an increasingly common component of biophysical and social-ecological system research in the Arctic. Projects on climate change, sea ice, coastal erosion, sustainability, and many other topics include some degree of interaction with and learning from local residents. Such interactions can take many forms, including the community workshop. The results of such workshops have been reported in many publications such as those just cited. Huntington et al. (2002) examined the structure of workshops in relation to exchanging information. However, little has been written about the interpretation of the discussions that take place in workshops. This paper attempts to address that gap, using examples from our experiences in Huslia, Alaska, on a project examining the human and ecological dimensions of wildfires.

A growing body of ecological and human dimensions research involves interdisciplinary collaborative projects that engage investigators from different academic disciplines in partnership with community members who have their own experiences and systems of knowing. Such collaboration offers substantial benefits, including enhancing the relevance of research with regard to local conditions and issues, providing opportunities for researchers to understand local and traditional knowledge, providing opportunities for local residents to learn from scientists, and developing conservation comanagement and other action plans that are beyond the reach of researchers, agencies, or community members alone.

Although it is generally agreed that cross-cultural and interdisciplinary dialogue can enhance the pool of available knowledge, experiences, and values from which insights can be drawn, it remains unclear as to how the cultural and personal perspectives of participants influence research outcomes. Whereas some perspectives are generally recognizable, such as differences in language, age, and gender, others are less apparent such as those reflecting different temporal and spatial perspectives, epistemological assumptions, or individual and factional motivations. Those cognitive and motivational differences can impede communication and lead to misinterpretation of research results, or they can enrich the research process, leading to mutual appreciation of the broader context.

In this insight paper, we highlight the importance of considering cultural, political, and epistemological context by reporting our experience analyzing data in an interdisciplinary study of the role of fire in affecting the resilience of Alaska Native communities and the relationship between wildfire and human activity in the boreal forest of Alaska and the Yukon Territory. Contrary to the experiences of Mosse (2001), we have found that careful attention to local knowledge, together with sensitivity to the way such knowledge is presented publicly, can effectively redefine the relationship between community members and visiting researchers. This in effect has enabled us to avoid what Cooke and Kothari (2001) refer to as the parochial dimension of participatory research. Last, our findings reinforce the conclusions drawn by Guijt and Shah (1998) in that effective cross-cultural collaboration demands considerable reflexivity throughout the research process, one aspect of which is to seek a diversity of types of interactions.

Our chapter begins with a description of the setting and the workshops themselves, including related interactions with community members. The discussion that follows addresses some of the complications we encountered while trying to interpret various statements made during the workshops, and the importance of understanding the context of such statements if one hopes to get the message. We conclude with observations on important elements in effective interpretation of workshop discussions, including a degree of caution on the part of the researchers.

BACKGROUND

Together with community partners, we are exploring the changing role of fire in the arctic-boreal climate system, particularly as it affects and is affected by human activities. This analysis includes human influences on the boreal fire regime such as ignition and suppression activities, and the effects of fire on people, including the availability of ecosystem services, e.g., production of game, berries, firewood, etc. We are also examining the role of wildfire-fighting employment in the economy of rural communities, for example, in relation to subsistence activities. This involves an examination of the short- and long-term consequences of current fire policy on northern interior communities.

The experiences described here are from a series of visits and workshops in the Koyukon Athabascan village of Huslia. In 2000, the community had 293 residents, of whom 274 were Alaska Native (U.S. Census Bureau 2000). It is located on the north bank of the Koyukuk River, approximately 467 km northwest of Fairbanks, within the Koyukuk National Wildlife Refuge. Huslia is in an area that is strongly affected by wildfire, although it is near the northwestern boundary of this fire-prone area. In 1949, the community was relocated from its former, frequently flooded site at Cutoff to the present-day location, an area known as Tsaatiyhdinaadakk'onh dinh or "where the forest fire burned the hill to the river" in Koyukon (Nelson 1983). Within a 32-km radius, large-scale fires burned in the late 1950s and 1970s. Aside from smoke inundation, Huslia was not directly impacted by the 2004 record-breaking Alaskan fire season.

A central figure in initiating and organizing these workshops and interpreting research results was research team member and co-author Orville Huntington, a Koyukon Athabascan, wildlife biologist with the U.S. Fish and Wildlife Service, vice-chair of the Alaska Native Science Commission, and long-time resident of Huslia. Co-author La'ona DeWilde, who was born and raised near Huslia and was at the time of the project a graduate student, was also a key part of data collection and interpretation. The other authors on this paper are nonlocal members of the research team and include Ph.D. scientists with expertise in ecology, anthropology, and interdisciplinary analysis of people and resources in the Arctic. In addition to interactions during the workshops and with local residents at other times, all members of the research team contributed to informal, preliminary interpretations of the workshop discussions, allowing us to refine our understanding and, in the case of evening conversations during our stay in Huslia, identify areas for further discussion on subsequent days, or even subsequent workshops.

In May 2003, H. Huntington and O. Huntington jointly organized and led a workshop in Huslia, specifically focused on observations of climate change. These discussions and the start of a project on wildfire led to a second workshop, in which all authors participated, held in Huslia in January 2004, and led by the two Huntingtons. This workshop also focused on local observations and traditional knowledge related to the impact of climate change in northern latitudes with an additional theme of the impact of climate change on wildfire and the impact of wildfire on plants, animals, and the community. More than 50 people of a range of ages attended this workshop from Huslia and neighboring Athabascan and Iñupiaq, i.e., Eskimo, villages, including roughly 20 high school students. A prominent feature of this workshop was the intentional educational interchange that occurred between village elders and youth. Both workshops followed the Koyukon practice of allowing elders precedence, yet opening the floor for anyone to speak. Neither workshop had a fixed agenda, but both involved presentations by visiting scientists as well as open discussions. In attendance at both workshops were climate scientists and the vice president for academic affairs and research at the University of Alaska Fairbanks. In each of these workshops, information was recorded in notes by the research team and as high quality digital audio recordings.

In March 2004, several members of the research team conducted semi-structured interviews with fire fighters and elders in Huslia and attended the Alaskan Western Regional Subsistence Advisory Board meeting held in this village. This board compiles feedback from villagers in the region and makes recommendations for fish and game management. Our final workshop was held in Huslia in October 2004, again led by the two Huntingtons, and recorded digitally. Discussions focused specifically on wildfire, including perceptions of fire, traditional knowledge of fire ecology and ecosystem services, cultural beliefs and practices regarding fire, local perspectives on fire management, and the social and economic role of employment fighting wildfires. This workshop was attended by about 10 elders and experts from Huslia and neighboring villages, including the Athabascan village of Nikolai, located on the Upper Kuskokwim River. Also in attendance at this workshop were federal fire management officers, one with the Bureau of

Land Management for the region and one for the Koyukuk National Wildlife Refuge, within which Huslia is located. Each workshop occurred over 2-3 d, and included opportunities for one-on-one discussions between researchers and workshop participants during breaks, shared pot-luck meals, and field trips.

The goal of these workshops was to engage local elders and others knowledgeable about fire in a collaborative, interactive, mutual-learning process with university scientists and researchers, community members, and representatives from fire management agencies (e.g., Huntington et al. 2002). The October 2004 workshop included a field trip to a nearby burned site. The workshops began with an extended period for introductions, giving participants a chance to learn about one another, and included long breaks, and community dinners. Visiting researchers then gave presentations, and the ensuing discussions wove in and out of several related topics, allowing for an organic progression rather than one dictated by a particular group of participants or by a preconceived idea of what was important or likely to be of greatest interest. Members of the research team periodically offered summaries, directed discussion, and requested clarification. Local elders also periodically summarized their perspectives, highlighting topics that they wanted to communicate to high school students who attended part of the workshops, or to participants from management agencies.

Our research goal was to collect traditional knowledge regarding wildfire impacts on communities and ecosystem services. Therefore, we deliberately invited knowledgeable elders and/or trappers with experience of fire effects to participate in the workshop. Statements presented here were made by village elders who, in accordance with Athabascan tradition, are afforded great respect and not contradicted in public. We did not aim to conduct community-wide opinion surveys and, therefore, do not report a range of variability in research findings. Made in any setting, isolated statements may on the surface appear ambiguous or extreme to outside researchers and even to local residents. In a public setting in a village such as Huslia, questioning or challenging such a statement by an elder would be considered improper. Even asking for clarification may imply disrespect. Thus, pursuing the meaning of such statements required not only tact but also the opportunity to discuss what was said with local residents, especially research team members, outside of the public setting.

Each community workshop and village visit was informative in and of itself. Taken together, these community visits and workshops allowed the research team, including community members, to triangulate the information they heard and to begin to put statements into a cultural and epistemological context. The series of visits and interactions also allowed visitors and locals to get to know one another, and familiarity produced greater trust and openness over time. Although it may not be possible to entirely remove ambiguity, placing statements in a larger cultural and economic context will result in more robust interpretation of data, as we discuss next.

DISCUSSION

Although some of what we heard during interviews and in workshops could be interpreted straightforwardly, other comments and statements seemed ambiguous or contradictory to the researchers. These apparent ambiguities underscore the importance of careful analysis and the interpretation of traditional indigenous knowledge within a larger political, spiritual, and epistemological context. As this case of human-wildfire interactions illustrates, the context in which fire is viewed has an important influence on perceptions of fire's impact and significance.

Our findings parallel those of Kendrick (2003). She emphasizes that the perceptions of government resource managers can be vastly different than those of indigenous resource users, regarding this as an opportunity for mutual learning and the development of innovative resource management processes. In her studies of Rangifer, caribou/reindeer comanagement in Alaska and northern Canada, Kendrick found that cross-cultural communication in studying social ecological systems that include indigenous people requires the development of mutual trust, which requires a degree of cross-cultural learning.

Worldview

On the coarsest scale of analysis, we can consider how different participants view wildfire. Western scientists and resource managers tend to take a reductionist approach; in this case isolating fire as the theme of analysis and organizing topic of research. While recognizing that fires are connected to wildlife habitat, employment, smoke inhalation, and so on, in the research process we consider these ecological and social connections not necessarily in and of themselves, but by virtue of their connection to fire. Fire and its ecological and social impacts are the primary concern.

In contrast, workshop participants from Huslia and other villages tended to take a much broader view, regarding fire as just one part of the larger ecological and socio-political setting. In our discussions of fire, for example, one elder mentioned land tenure, land status, and lack of local control of resources on at least two occasions. She summed up her feelings at one point by saying, "We need to follow Indian law, instead of always getting pushed around by Europeans." In this context, local inability to control fire policy on federal and state land, despite being affected by the consequences, was not an isolated phenomenon, but another instance among many in which local people lost the power to run their own lives (Berger 1985).

Similarly, a question by researchers regarding the effect of fire on moose habitat led to a discussion of how low water levels in the rivers and a different pattern of melting in recent years have resulted in less ice scour along the riverbanks. The ice scour had been a source of disturbance, helping create good moose browse, but now other sources of disturbance are needed, such as flood or fire, to create productive habitat. For the Athabascans living subsistence lifestyles, fire was not the central concern, but one of many ecological, social, and political phenomena that impact their lives.

"We Hate Fire": the Political and Interpersonal Contexts of Fire

In science, unexpected or surprising results are particularly interesting as they may indicate a need for theoretical or methodological innovation. In our workshops in Huslia, statements that surprised researchers were particularly challenging to interpret, requiring careful contextual consideration. The most striking example occurred when, in response to an open-ended inquiry about the economic and social impacts of fire fighting employment in rural communities, a respected elder said, "We hate fire." Members of the research team, including the locally resident researchers, were surprised to hear such a vehement and extreme statement, particularly as hatred is viewed as inappropriate in Koyukon culture.

At the time, the visiting researchers considered the political context to find meaning in this surprising statement. The elder's vehemence about fire may have been tied to her frustration about land tenure and ownership, and the inability of local residents to participate meaningfully in decisions about fire management on federal lands, that influence their livelihoods. As noted in the Appendix, these fires on federal lands sometimes burn shelter cabins, upon which local people rely for safe winter travel, and trapping cabins, which local residents or their ancestors built long before land-tenure status was legally formalized in the region.

Representatives from federal agencies that implement national fire policy and make decisions about fire management on lands that local people rely on for subsistence were at the workshop. When analyzed within the political context, this elder appeared to view our workshop as a platform to address managers about fire policy, rather than as an event specific to our research project.

Although analysis of this surprising statement in a political context seems plausible and reasonable, different interpretations of the statement's context further reveal the complexity of studying traditional indigenous knowledge. Careful analysis of the workshop's audio recording revealed a pause before the word "hate." Although the elder communicates fluently in English, it is not her first language. Visiting researchers wondered if this pause might indicate that she was searching for the right word to use, and that perhaps in the translation the connotations carried in English by the word "hate" were not shared in Denaaga, her first language. Instead, O. Huntington explained that such a pause is a way to add emphasis, to signal to the audience that something important is about to be said. The potential for divergent interpretations is clearly large.

Finally, we cannot overlook the social context of this statement. Within the two previous years, two community members had died in a cabin fire in Huslia. In addition, during the course of our workshop, there was a fire in one of the cabins in Huslia. Although no one was hurt, the collective effort of many community members was required to extinguish the fire. Although we are not certain of the total context of this elders' statements, it seems clear that interpreting her statement, strictly as an ecological one, i.e., that fire serves no useful function on the landscape, would have been a mistake.

"Some Things you don't Talk about": the Spiritual and Cultural Context of Fire

In the Koyukon worldview, the spoken word has great power. It is often inappropriate to speak of sentient forces, and it is bad luck to speak ill of them, lest they hear and take offense. At times it is inappropriate to speak of sentient forces at all. Fire is one such force, which in part explains our surprise at the negative statement discussed above. As we detail below, several aspects of the spiritual power of fire were discussed, but most notable was what could not be discussed. On more than one occasion, elders and respected community members responded to researchers' inquiries with the comment, "Some things you don't talk about." At one point there was an extended discussion about the personal struggles that elders and community members have experienced. Although they know that it is culturally forbidden to speak of a certain animal, phenomenon, or situation, they are compelled to do so to represent and defend Native perspectives on resource management issues that are dominated by western perspectives and procedures.

One elder told us of how she had been instructed not to speak of what a fire might do because speaking in that way might cause the fire to do exactly as she had said. Several people talked about using ashes to protect the food that is stored underground, because insects, mice, and other animals would not go through the layer of ash. Although this practice may have a physical basis, the same principle has a spiritual dimension as well, as seen in the practice of sprinkling ash around a house to provide protection from earthquakes and other misfortunes.

With the power of words and of fire in mind, a question that was asked of the scientists by a local resident about the effect of ash on fish takes on an additional dimension. There is a biochemical aspect to it, which scientists could address, but there is also a spiritual aspect regarding the way fire transmits its power through ash to water and across the landscape. The person who asked the question may have appreciated an answer about the biochemical effects had we been able to provide one, but may not have been completely satisfied with that answer. Similarly, the protection afforded by ashes to buried food may be recognized as spiritual as well as biochemical. Our initial inclination to identify only a biophysical explanation may have missed part of the point of the conversation.

"It all Started when they Put a Man on the Moon": Epistemological Context

Perhaps the most challenging aspect of partnering with indigenous communities and linking traditional ecological knowledge with western scientific knowledge is accounting for the diverse ways in which people see the world and the accompanying diverse beliefs about how we come to gain knowledge about the world. As this example illustrates, the ontology, i.e., how we see the world, and epistemology, i.e., how we gain knowledge about the world, of Native Alaskans differs considerably from that of western trained scientists. A true partnership requires mutual respect and the ability to create bridges between these differing worldviews.

A particular focus of the first workshop was local observations of climate change, which impacts northern latitudes acutely and is a topic of considerable interest in Huslia and in many other Alaskan villages. Although villagers' observations of the many local environmental changes occurring in recent years are consistent with scientific observations and findings around the state and the Arctic, some of their explanations of these changes are not. On several occasions, local participants described how a respected leader had predicted that substantial social and environmental upheaval would result from the Apollo moon missions (1969-1975). The moon has considerable power and significance in Koyukon cosmology. Tampering with the moon in the form of lunar space landings was disrespectful and inappropriate and could only lead to negative repercussions. DeWilde notes that there may be intergenerational differences, too, among the Koyukon, with younger generations having the same general perspective but perhaps adding physical explanations to spiritual ones. Because this prediction was made at the time of the moon landings and has been followed by climate and other environmental changes, the prediction has become an explanation.

The phenomenon of global warming and the potential role of humans are well known to Huslia residents. The moon landings do not simply fill a void left by the absence of other explanations. Indeed, the moon landings have a symbolic role, too, in relation to humankind's pursuit of technological and other forms of progress without careful consideration of the consequences: landing on the moon was a boastful and arrogant thing to do to another spirit. Nonetheless, it can be difficult to reconcile the Apollo explanation with current scientific understanding of climate. Some visiting researchers, though not the climatologists, speculated on the role of rocket fuel and vapor in the stratosphere, seeking a physical explanation to connect the two. Others made the symbolic interpretation already mentioned, whereas some preferred to move on to another topic.

In this example, the invocation of Koyukon cosmology created a mismatch in the worldview between non-Native western scientists and local Native researchers and residents. This mismatch posed a challenge to communication and the sharing of ideas. It is unlikely that the local participants expected the visiting researchers to fully accept the idea that the moon landings were responsible for environmental disruption. At the same time, the fact that they mentioned the idea at all, and on more than one occasion, indicates they were trying to convey a point. The disruption caused by the modern way of life often arises in conversations with Alaska Natives. Although modern conveniences are appreciated, there is considerable concern for maintaining traditional language, stories, customs, and knowledge, and the social and other problems of being marginalized in the western world of today are all too real. The disasters that follow hubris are found across many cultures, and so the symbolic interpretation of the moon landings is no doubt part of what we were being told.

It is also likely that, whether we were expected to believe it or not, the local participants also wanted us to understand and respect the strong connections they see between human thought and action and the natural world that guide their interactions with the world around them. In this context, management actions such as prescribed fires have much broader implications for interactions between the community and the natural world than might be apparent to fire managers and researchers.

For the visiting researchers, the discussion about the moon landings could also be viewed as something of a test. Our reactions may have partly determined the degree to which the local participants wanted to continue to help us or take part in the discussion. Had we simply replied, "That's ridiculous," the workshop might have effectively ended. Had we sought purely physical explanations, we might have been viewed as slow learners with a limited worldview. Had we accepted the explanation completely and abandoned other lines of cause or inquiry, we would probably have been laughed at. Huslia residents have their own expectations of how researchers work and think. We would at least like to think that the visitors' acknowledgment of the significance of the prediction and its importance in local cosmology helped make participants more comfortable discussing additional aspects of their worldview without fear of ridicule.

CONCLUSION

In interdisciplinary, collaborative community workshops such as those described here, the challenge for each researcher or other participant is to identify and interpret the information of relevance without losing sight of the larger political, spiritual, and epistemological contexts in which that information is presented and understood. We are reluctant to offer prescriptive steps or to claim to have a fail-safe method for achieving this goal. Our experience does, however, suggest some useful approaches to keep in mind, as well as the need for caution on the part of the research team, including a degree of humility in recognizing that complete understanding is elusive.

One approach to capturing discussions appropriately is to begin the process of analysis and interpretation during the workshop itself so that those steps become part of the collaborative effort. In our most recent workshop in Huslia, we spent the final morning presenting to workshop participants and discussing tentative conclusions and papers that we might write, including this one. Another approach to help retain context is to take time and, ideally, to have more than one workshop or visit to the community. Community-researcher connections develop over time and need many opportunities to interact in different settings. The workshop, which included several breaks that often extended for half an hour or more, allowed for much informal interaction. The second workshop was accompanied by a field trip. We had evening gatherings, which provided a social setting in which people could better get to know one another.

A single workshop can achieve a great deal, but repeated events allow relationships to develop further, trust to grow, and demonstrate a higher level of commitment to the shared enterprise. Furthermore, variety in the types of interactions can help provide opportunities to hear from different people and in different contexts, to learn more about local context, and to obtain a broader range of information to help in overall interpretation.

The composition of the research team is clearly another key element of workshops. Having local residents as part of the research team greatly enhanced our introduction to the community as well as our ability to review and discuss what we had learned. Anthropological expertise and experience was also valuable, although most members of the research team had extensive experience working in rural or Native communities prior to the Huslia workshops. Those with limited experience in this regard benefited from some informal coaching prior to and during the workshops. Although neither local researchers nor anthropologists are perhaps strictly necessary for a successful workshop or similar interaction in a community, their skills and insights are important ingredients that need to be included in some fashion.

To truly listen to and understand traditional ecological knowledge, researchers must ' e able to navigate across the epistemological and practical differences and to recognize statements within their larger historical, political, spiritual, and epistemological contexts. Researchers and agency personnel tend to have a very narrow focus with specific goals, in this case specific to the impact of fire on the social-ecological system. However, the villagers have a much broader perspective on the entire social-ecological system, including its history and political landscape. When they speak of fire, they are also discussing the larger political, spiritual, epistemological settings of their lives and their culture. These topics are not merely academic interests, but vital to their daily lives. Respecting that larger context is a crucial element of success in community workshops.

5

A Vista for Microbial Ecology and Environmental Biotechnology

Microbial ecology and environmental biotechnology are blossoming fields that are taking advantage of profound advances in biology, materials, computing, and engineering. Although traditionally microbial ecol-ogy and environmental biotechnology have been distinct disciplines, their futures are intimately linked. Together, they offer much promise for helping so-ciety deal with some of its greatest challenges in environmental quality, sustainability, security, and human health. ly, what are the interrelationships among the com-munity's members and between them and their environment? Interrelationships involve spatial or-ganization (i.e., who is near whom) and materials that the microorganisms exchange. Understanding these interrelationships is the ultimate goal of mi-crobial ecology.

Answers to these four questions usually change over time and especially in response to perturba-tions to the environment. The ability of microbial communities to respond to environmental changes (natural or anthropogenic) is called community re-silience and stability, and it influences each of the four questions.

The beginnings of microbial ecology can be traced back at least to the late 1940s and 1950s; great conceptual advancements were made in the 1960s and early 1970s. However, microbial ecology struggled as a scientific discipline for decades be-cause of a simple reality: Tools to address the four fundamental questions either did not exist at all or were unreliable.

The small sizes and simple shapes of microorganisms made morphology a tedious and insufficient gauge of identity. Selective culturing on the basis of metabolic function was a giant step for-ward, conceptually and in practice, but often failed or gave biased results. Thus, brilliant microbiolo-gists labored to gain the faintest glimpse of the di-versity and organization of microbial communities.

They knew that so much more could be learned, if they only had the tools.

The use of molecular biology tools, beginning around 1985, started to change the prospects for microbial ecologists. Selective and reliable ampli-fication of defined DNA with the polymerase chain reaction (PCR) and hybridization with DNA oligo-nucleotides made it possible to interrogate directly the genetic information of individual microorgan-isms and entire communities.

The small-subunit ri-bosomal RNA (SSU rRNA, also known as 16S rRNA for prokaryotes) was the first target for hybridization.

What are these exciting fields? How do they con-nect to each other? And what is their common vista?

How will microbial ecologists and environmental biotechnologists invest their resources so that the two fields, working as partners, create the greatest scientific advancements and benefits to society?

That was the question for the symposium held last April at Arizona State University. The symposium's 10 speakers, who are also the authors of this arti-cle, are all considered international leaders working at the interface of microbial ecology and environ-mental biotechnology. These experts participated in focused discussion sessions aimed at arriving at a consensus view on the most important directions for the partnership of the two fields.

This Viewpoint describes the dramatic and promising vista that the symposium's participants saw for the partnership of microbial ecology and environmental biotechnology.

MICROBIAL ECOLOGY—THE SCIENCE

Microbial ecology is a long-standing scientific dis-cipline that is undergoing remarkable, even revo-lutionary, changes. The core of the field aims to understand microbial communities, which are self-organizing and self-sustaining assemblages of dif-ferentmicroorganisms,andhowthesecommunities interact with their environment.

The science of microbial ecology tries to an-swer four fundamental questions. First, what mi-croorganisms are present in the community? The phylogenetic makeup (identity and number) of the microorganisms present in a community is called a community structure. Next, what are the capabili-tiesofthemicroorganismsforcarryingoutreactions that transform the community's environment? The catalog of catalytic capabilities is called the com-munity's phenotypic potential. Third, what reactions and transformations are the community members actually performing? The realization of phenotypic potential is called the community's function. Final-and is still the most widely used. This rRNA gives information on the phylogenetic identity of the microorganisms, and this addresses the first fun-damental question of microbial ecology.

Other targets within the genetic system of micro-organisms answer different questions. For example, amplifying and detecting specific genes in the genome of microorganisms in the community defines the phenotypic potential, and this addresses the second question about the capability of microorganisms. Messenger RNA (mRNA) reveals which genes are being expressed and, therefore, which functional proteins like lyare formed.This gives information on the expressed phenotypic potential, or the third question about the community's function.

Assessing RNA targets by using microscopic visualization, such as with fluorescence in situ hybridization (FISH), adds information on spatial organization. Recently, microbial ecologists have begun to clone and sequence large genome fragments from various microbial communities; therefore, it is now possible, at least for low-diversity communities, to reconstruct entire genomes of uncultured community members. In addition, microauto-radiography, coupled with FISH (i.e. ,MAR-FISH) and stable-isotope probing, makes it possible to detect the function in the community, even when the researcher has no idea what genes or mRNA should be detected.

Environmental biotechnology-The service Environmental biotechnology manages microbial communities that provide services to society. Prominent and emerging services include removing contaminants from water, waste water, sludge, sediment, or soil; capturing valuable products from renewable resources (e.g., biomass), particularly energy carriers but also nutrients, metals, and water; sensing contaminants or pathogens in the environment or, perhaps, in humans; and protecting the public from dangerous exposure to pathogens.

These services are essential if modern human society is to be safe, sustainable, and secure. Microbiological communities, properly managed, can provide these services reliably, continuously, economically, and without creating other hazards. Environmental biotechnology is almost a century old, although this name is relatively new. Past names include biological treatment, biological processes, bioprocess engineering, bioremediation, and bioenvironmental systems. The new name reflects that environmental biotechnology is adapting and benefiting from the modern tools of molecular biology, as well as from other advances in science and technology. The scientific foundation of environmental biotechnology is microbial ecology. Because environ-mental biotechnology ultimately aims to manage microbial communities for the good of society, a deep understanding of microbial communities-that is, microbial ecology is essential. Environmental biotechnology addresses real-world goals by managing microbial communities, rather than finding or creating the solves-all-problems "superbug".

The box at left lists criteria for a successful en-vironmental biotechnology for removing contami-nants or resource capture.

These criteria demand a robust, self-organizing, and self-sustaining microbial community; they do not cater to a highly specialized or pampered superbug. Thus, the goal is to understand microbial communities so that the system's conditions lead to robust communities that provide the desired service.

NEEDS PULL, SCIENCE PUSHES

A field that has been around for a century usually does not show a burst of creativity, but environ-mental biotechnology is doing just that today. The reason is the convergence of a strong "needs pull" with a strong "science push". Today, environmental biotechnology gets a powerful pull from the needs of human society arising from increasing challenges to achieve sustainability, environmental quality, se-curity, and human health.

The sustainability of modern human society depends on extracting essential materials from renewable resources and reducing reliance on non-renewable resources. At the top of the list of at-risk resourcesaretwothatenvironmentalbiotechnology addresses directly: water and energy. Today, water providers are forced to tap sources of lower quality:

polluted ground and river water, eutrophic lake and reservoir water, and wastewater. These poor-qual-ity sources need substantial treatment to eliminate public-health risks, unpleasant taste and odor, and discolora-tion. Experts on the Middle East predict thatwater-notoil-willbethegreatestfuturecause of strife in this re-gion.

After water, energy is the most precious resource, and future sources must shift from fossil fuels to renewable ones. Environmental biotechnology is at the heart of upgrading poor water sources for human use and for converting renewable energy sources-particularly biomass and sunlight-to useful forms, including natural gas, hydrogen, and electricity. Organics, nutrients, and metals that are not captured but are instead discharged to the environment become pollutants, not resources.

Environmental biotechnology has a long-standing role in treating wastewater and other contaminated water, air, and solids. Increasing population, urbanization, and economic activity heighten the need to apply environmental biotechnology to preserve (or im-prove) environmental quality, along with capturing valuable resources. Innovative environmental biotechnology approaches appear to be well suited for improving environmental quality in developed and developing countries alike. The readiness of developing countries to use these technologies lies in their improved reliability and operability in a decentralized setting.

Criteria for a Successful Environmental Biotechnology for Detoxification or Resource Capture

The technology must be practical at a large scale. rel-evant units normally are thousands of cubic meters per day for flow or tons for mass.

It must operate reliably and continuously around the clock. The concept of a "time-out" normally is not relevant. The technology must be relatively simple to operate and largely self-correcting without much hu-man intervention. It must be economical to build and operate. The services provided are so essential to society that they cannot be priced out of the reach of those on the low-est rung of society's economic ladder. Security and human health also relate to envi-ronmental biotechnology. Infectious diseases from pathogenic microorganisms remain the main cause of death worldwide, and terrorist threats to dissemi-nate pathogens heighten the danger. Microbial systems can monitor for pathogens or chemical toxins in hospitals, water supplies, the air, and, perhaps, even humans. Microbial systems also contribute to cures or therapies by producing drugs or enzymes to fight diseases.

Environmental biotechnology can contribute in all these ways because microbial communities have seemingly infinite ways to live, even in envi-ronments that appear to be bizarre or hostile. Their ability to organize and sustain themselves provides human society with a cornucopia of metabolic potential to find services to benefit society. Microbial ecology, the core scientific discipline, allows us to understand microorganisms as part of their com-munities:"to think like the microorganisms." Armed with this deep insight, the environmental microbi-ologist can create sustainable systems that "work for the microorganisms so that they work for us." On the science side, the great advances in mi-crobial ecology, previously summarized, stand at the head of the science push and tend to garner the greatest credit. However, microbial ecology is not the only strong science push behind environmental biotechnology at the beginning of the 21st century. The products of modern materials science provide another hearty push, perhaps as important as from microbial ecology. The materials push began around 1970, when lightweight, high-strength plastics made possible biological towers for wastewater treatment, which were the first biofilm processes with high surface areas and small footprints.

From the 1980s through the 1990s, lightweight bio-film carriers in the form of gravel-sized pellets made even more compact, high-rate processes possible. From the late 1990s to today, microfiltration membranes have been replacing gravity separators for activated sludge, improving effluent quality, re-liability, and compactness. More recently, the membrane biofilm reactor makes it possible to use H_2 toreduceNO_3-,ClO_4-,andalargerangeofoxidized contaminants in drinking water, groundwater, and wastewater. Recent advancements in nanoscale (1-100-nm) materials and biomicro-electromechanical systems(bioMEMS) technologies surely will provide similar opportunities.

Another science push comes from mathematical modeling, the ultimate tool for integrating the large number of microbiological, chemical, and physical processes that occur in any microbial community of environmental significance. A model uses mass-balance equations to represent the significant com-ponents in the community. Creating a model demands that the modeler identify the important system components, for example, the critical types of microorganisms, the substrates they consume, and the products they produce. Likewise, the mod-eler must represent the important reactions, such as the synthesis of new biomass, consumption of substrates, and generation of products, with math-ematical expressions that capture what is known about the microorganisms and how they function.

On the one hand, modeling is an exercise in integrating all that we know about microorganisms so that we can test and exploit our understanding of how the microbial communities work as a functioning ecosystem. On the other hand, modeling is the best way to find out what we do not understand about the microorganisms and their interrelationships with each other and their environment. When we cannot model a system, we can identify what we do not understand, and this helps us direct fundamental research to the most important questions.

Whether the goal is gaining scientific under-standing or applying that understanding to create high-value services or products, the most funda-mental questions are at the interface of microbial ecology and environmental biotechnology. Perhaps the overarching question is, "How do microbial communities self-assemble to achieve and maintain a function?" The environmental biotechnologist depends on self-assembly once the proper conditions are in place. Therefore, self-assembly needs to be based on identifiable principles that can be put into prac-tice. The overarching question then leads to corollary questions: What are the underlying principles of self-assembly? What community structure is optimal for a biotechnological application? What environmental conditions trigger assembly of the desired community with its desired function? Are the conditions and outcomes predictable, reproducible, and controllable? Self-assembly also implies that microbial communities behave as a kind of multicellular organism, one in which the whole is more than the sum of its parts. Corollary questions along this line of inquiry include: Does the community have a multicellular A field that has been around for a century usually does not show a burst of creativity, but environmental biotechnology is doing just that today.

"life of its own"? How do the microorganisms communicate and coordinate to gain the advantages of a multicellular state? What are the advantages, if any, to the community and to humans?

A THREE PEAK VISTA

The consensus vista for the partnership of microbi-al ecology and environmental biotechnology shows three "peaks" in the distance, representing different themes: more powerful analytical tools, integrated "-omics" approaches, and research that is more theory-or process-driven.

The strongest science push in microbial ecology comes from the explosion of new molecular tools. These tools have propelled the field forward, which makes it possible to answer questions that we could not think to ask in the not-so-distant past. Nonethe-less, existing tools are inadequate for studying the complexity of microbial ecosystems and supporting the design and operation of innovative environmental biotechnology approaches. Shortcomings include methods that produce results too slowly and with too much effort, have biases, are too expensive, offer insufficient quantification, and lack coverage over the ranges of structure and function that are impor-tant in relevant microbial communities. To overcome these shortcomings, developers of new molecular tools need to meet four criteria. First, the methods must be high-throughput, generating relevant data in minutes to a few hours. Second, the outputs must be quantitative enough to provide the information that sorts out the "lead actors" from the "bit players" for a given function and sufficiently sensitive to find the important microorganisms or reactions even when others dominate in numbers.

Third, the outputs must provide the type of infor-mation that reveals structure and function in paral-lel, because they are intimately connected. Finally, methods must put more emphasis on the structure and function of eukaryotes and phages in order to yield a more complete picture of community struc-ture and function. Ideally, more than one method should be applied to generate data (e.g., PCR-based and PCR-independent approaches).

Integrate the "-omics" approaches. Most mo-lecular interrogation has been directed toward DNA, focused on selected genes, or (recently) aimed at high-throughput genomics. Although the potential for expanding capabilities in environmental genomics is enormous and must be pursued with vigor, resources also need to be in-vested in developing and using tools on the basis of the other molecules within the cells, or the other-omics disciplines.

The primary other-omics are transcriptomics, which is the study of mRNA as an indicator of gene expression; proteomics, which fo-cuses on protein (enzyme) identification, characterization, and quantification as they relate to cellular function; lipomics, which targets the study of mem-brane lipids; and metabolomics, which studies the metabolic intermediates of cellular functions. Each of the several-omics approaches provides different but complementary information that re-veals distinct information about communities. Information from genomic and proteomic stud-ies forms a comprehensive picture of the cellular response to perturbations or other changes in the environment; however, metabolomics reveals the ultimate functional response and is most likely to predict community phenotype.

A great challenge is to borrow strategies that were developed for the analysis of single organisms and apply them to complex microbial communities containing many and often unknown species. Ini-tial success stories of monitoring gene expression and protein formation are encouraging and provide important guidance for future designs. The hallmark of high-value-omics technology is the integration of the high-throughput data. An essentialelementisbioinformaticstools,whichoffer database support for data management and mining and biostatistics to maximize the benefit of these expensive and labor-intensive methods. Our current format for disseminating scientific results (e.g., refereed journal papers) is not ideally suited to handle the massive data sets of-omics-based experiments.

Long-term stewardship and maintenance of these data over time must be addressed.

COMPLETE BIODEGRADATION OF PCE

Tetrachloroethene (PCe) and trichloroethene (TCe) are all-too-com-mon groundwater contaminants that threaten human health. Ba-sic research over the past decade brought to light a novel group of microbes, the Dehalococcoides, that can detoxify the troublesome chloroethenes. Dehalococcoides capture energy from reduc-tive dechlorination reactions, thus making a living by reducing toxic groundwater contaminants. The reductive dechlorination pathway that leads to the complete detoxification of PCe and TCe to nonhaz-ardous ethane (eTH) is shown below. a critical aspect is removing all chlorine substituents from the ethene backbone, because the last chlorinated intermediate, vinyl chloride (VC), is more toxic than the parent compounds. some Dehalo-coccoides organisms, such as strain BaV1, efficiently detoxify dichlo-roethene (dCe) and VC, thereby producing environmentally benign eTH and inorganic chloride.

Consortia containing Dehalococcoides organisms have been successfully applied at field sites where the intrinsic microbiol-ogy was insufficient to sustain acceptable degradation rates. Knowing the key players (i.e., Dehalococcoides) and understanding the microorganisms' ecology were critical for successful technol-ogy implementation. For example, nucleic-acid-based tools (e.g., quantitative real-time [qrTm] PCr) were designed to specifically detect and quantify Dehalococcoides 16s rrna genes and genes implicated in VC reductive dechlorination. These diag-nostic and prognostic tools changed chloroethene bioremediation to a predictable science with obvious benefits to society.

Make the research more theory-or process-driven. Molecular-based research in microbial ecology has developed an image of "stamp collecting", or the gathering of a lot of data because it is possible with the available tools. This enthusiasm for data collection is natural in an emerging field and was necessary when so little information was available before. Having any information is a giant leap for-ward, and no one knows a priori what will be the most interesting findings. Therefore, collecting microbiological stamps has been essential and will never disappear.

After 15 or so years of profitable stamp collecting, microbial ecology and environmental biotech-nology are poised to advance to a higher plane of molecular-based research. This next step will em-phasize research driven by ecological theory, pro-cess needs, or both. The field of macroecology has developed elaborate theories about ecosystem succession, diversity, invasion, and stability, such as the hypothesis that highly diverse ecosystems are more stable. Statistical and deterministic modeling tools have been developed to design experiments and interpret results. For the most part, these theories have not penetrated microbial ecology.

Without a doubt, the application of ecological theory from macroecology will demand a major modification to take into ac-countthefundamentaldifferencesbetweensystems of macroecology and microbial ecology. For ex-ample, microbial ecosystems deal with huge numbers of cells (e.g., more than trillions of bacteria per liter of water) that often cannot be differentiated as individuals; this contrasts with much smaller numbers of identifiable individuals in a macroecosystem. In addition, macroecosystems are based to a larged e-greeon specific prey- predator relationships, whereas microbial ecosystems more often target the parallel metabolism and exchange of materials, despite predation being important (albeit poorly understood). Thus, the simplistic application of macroecology theory to microbial ecosystems would likely back-fire. Nonetheless, proper application of ecological concepts-such as diversity, stability, competition, redundancy, and allied tools-provides a powerful intellectual framework for designing research and understanding the meaning of the results.

The next plane of research also needs to be pro-cess-driven rather than tools-driven. Now that a wide array of tools is available and continuing to expand, the researcher can ask critical questions about the process and then pick the right tools to find the answers. This approach allows researchers to take a more holistic approach toward understanding com-munity structure and function. The researcher can form a hypothesis and bring in a range of molecular, modeling, and other tools to gain information that tests the hypothesis.

A powerful research tool is to perturb the microbial community and then observe whether the organisms respond in the predict-ed manner. In this hypothesis-driven approach, the questions lead to selection of the tools; this is a wel-come reversal from the common situation in which tools dictate what questions can be addressed. Theory-and process-driven research must rec-ognize the role of scale-spatial and temporal. Microorganisms are generally on the order of a few micrometers in size, and microbial aggregates are only occasionally >1 mm. The technological pro-cesses in which microbial communities are used are very much larger. Even a bench-scale bioreactor has dimensions in centimeters, and full-scale systems measure in meters, often in tens of meters. Research that focuses on the micrometer-to-millimeter scale relates to what happens at the scale of meters, but the application is not always simple.

BIOELECTRICITY FROM THE MICROBIAL FUEL CELL

More than 20 years ago, researchers noticed that a fuel cell containing microorganisms could generate electricity, but how it occurred was a mystery. several years later, researchers started dis-covering that certain bacteria could transfer electrons to solid-phase acceptors, such as iron oxides. If they could reduce a solid-phase acceptor, could they also transfer electrons to a solid-phase electrode acting as an anode in a fuel cell? The answer was a resounding "yes". Thus, the revolutionary concept of a microbial fuel cell (MFC) was born. as shown in the diagram, an MFC operates in the same man-ner as does a conventional fuel cell, except that bacteria living as a biofilm on the anode catalyze the oxidation of the fuel.

The MFC is still at an early research stage, but it can be a revolutionary breakthrough for capturing renewable energy sources. The bacte-rial catalysis at the anode makes it possible, for the first time, to use renewable organic material (biomass) as a fuel for fuel cells. In contrast, conventional fuels cells use an expensive platinum catalyst on the anode and still can use only high-grade H_2, which is obtained by reforming petroleum, a nonrenewable energy re-source. By operating at ambient temperature and without com-bustion, an MFC can more than double energy-capture efficiency while eliminating air pollution. recent work shows that specialized bacteria are not necessary to generate electricity from organics in an MFC, but inspiration comes from the fundamental studies on physiology, ecology, and genomics of particular metal-reducing bacteria.

Molecular-based Research in Microbial Ecology has Developed an Image of "Stamp Collecting".

In such a situation, modeling can be of special value, because it takes into account transport processes and heterogeneity. In parallel with spatial scales, the range of temporal scales is extreme. A classic microbiology experiment might last 1 day, and an ex-periment on microbial ecology might last for a few months and have controlled operating conditions; however, a real-world process operates continuously for years and with unpredictable changes in temperature, substrate loading, and other factors. It seems logical that increasing time horizons and variabil-ity affect community structure and, perhaps, function. These effects need to be addressed directly in research design and how results are interpreted.

Where are the Benefits and Roadblocks?

Skeptics demanding proof that "molecular research is really improving environmental technology" have confronted all of the participants in the vista symposium. Skeptics want specific examples of cause and effect in which research led directly to a new or improved process. Perhaps they seek a single "magic bullet". Up to now, direct, incontrovertible proof has been scarce, but it is beginning to emerge.

The demand to see cause and effect is justifiable, and the partnership of microbial ecology and environmental biotechnology needs to satisfy the demand. Achieving the three-peak vista will go a long way toward satisfying critics and attracting new supporters of the partnership. The immediate and direct payout from the science research is thinking like the micro-organisms.Then, good engineering and modern materials can translate the understanding into systems that manage the microbial communities to provide new and better services: working for the microorganisms so that they work for us. It is a two-part process that grows and harvests the fruit of microbial ecology and environmental biotechnology. Although the partnership of microbial ecology and environmental biotechnology may not have such dramatic results yet, theboxesonpages1100-1102 give concrete examples that fruits are beginning to be harvested.

Despite these impressive achievements that are beginning to emerge and the high potential for the partnership, road blocks are inevitable because of the inherent complexities of environmental microbial ecosystems.

First, the numbers of different microbial strains are enormous-de facto infinity. Further-more, only a tiny fraction of the strains have been cultured and characterized.Second, microorganisms can evolve rapidly-we face a perennial "movingtar-get".

Third, environmental microbial ecosystems are physically complex aggregates that change in time and space.

Fourth, the environmental matrices of sludge, wastewater, sediment, and soil are "dirty" -a special problem for extracting cellular material and avoiding interferences.

Fifth, a microbial ecosystem needs to be understood in terms of its structure, function, and interactions with other ecosystems in order to achieve society's goals.

In summary, the study of microbial ecology in the context of environmental biotechnology is substantially more difficult than its study in medical settings, where many of the tools originated. Thus, connecting microbial ecology to environmental biotechnology always will demand a lot of hard work, persistence, and creativity as road-blocks are identified and overcome. Modern-day microbial ecology and environmen-tal biotechnology are young disciplines, still finding their identities separately and as partners.

As with any emerging field, its researchers and the society that supports them must have a degree of faith that an investment of resources will yield a large benefit, because this is the way breakthrough advances in science and technology often have been made. Sci-entists first construct a "cathedral of knowledge", which inspires and enables creativity that leads to the great technological breakthroughs. Perhaps mi-crobial ecology and environmental biotechnology are putting up the buttresses of the cathedral now.

Detecting Airborne Infectious Diseases

The recognition of emerging airborne infectious diseases, such as avian influenza and severe acute respiratory syndrome (sars), as well as heightened concerns over bioterrorism, has put a spot-light on the persistent public-health problem of airborne infectious disease. For decades, public-health professionals were poorly equipped to measure or identify pathogens in air. now, the applica-tion of microbial ecology principles and techniques to aerosols holds tremendous promise for explaining airborne disease epidemiology and designing systems that minimize our exposure to allergens, cold viruses, and other current and emerging infectious airborne agents. For example, although >1.5 million people die from tuberculo-sis each year and one-third of the world's population has been exposed to Mycobacterium tuberculosis, viable bacteria have never been cultured from an ambient air sample. M. tuberculosis cells were recently detected in indoor air by a polymerase chain reaction (PCr) assay. PCr-based measurement of rhinoviruses in office-build-ing air, in conjunction with human epidemiology stud-ies and measurement of ventilation rates, has led to a rethinking of how humans catch colds. The photograph shows a dna microarray with the potential for describing >9000 taxonomic groups.

The study of microbial ecology in the context of environmental biotechnology is substantially more difficult than its study in medical settings.

6

Environmental Impact Assessment

An Environmental Impact Assessment (EIA) is an assessment of the likely positive and/or negative influence a project may have on the environment. "Environmental Impact Assessment can be defined as: The process of identifying, predicting, evaluating and mitigating the biophysical, social, and other relevant effects of development proposals prior to major decisions being taken and commitments made." The purpose of the assessment is to ensure that decision-makers consider environmental impacts before deciding whether to proceed with new projects.

OVERVIEW

The US Environmental Protection Agency pioneered the use of pathway analysis to determine the likely human health impact of environmental factors. The technology for performing such analysis is properly labelled environmental science. The principal phenomena or pathways of impact are: soil contamination impacts, air pollution impacts, noise health effects, ecology impacts including endangered species assessment, geological hazards assessment and water pollution impacts. Pathway analysis and The Natural Step definitions subsequently became the basis of the global ISO 14000 series of environmental management standards and the more recent ISO 19011 auditing standard; however, these ISO standards are not in common use in the U.S. and most other countries.

After an EIA analysis, the Precautionary Principle and Polluter Pays may be applied to prevent, limit, or require strict liability or insurance coverage to a project, based on its likely harms.

Environmental impact analysis is sometimes controversial and contested. Related analysis of social impacts is achieved by Social impact assessment. Analysis of business impacts is achieved by context analysis. Design impacts are assessed in relation to context theory. EIA predicts what a specific action can do to the environment.

EIA AROUND THE WORLD

China

The Environmental Impact Assessment Law (EIA Law) requires an environmental impact assessment to be completed prior to project construction. However, if a developer completely ignores this requirement and builds a project without submitting an environmental impact statement, the only penalty is that the environmental protection bureau (EPB) may require the developer to do a make-up environmental assessment. If the developer does not complete this make-up assessment within the designated time, only then is the EPB authorized to fine the developer. Even so, the possible fine is capped at a maximum of about US$25,000, a fraction of the overall cost of most major projects. The lack of more stringent enforcement mechanisms has resulted in a significant percentage of projects not completing legally required environmental impact assessments prior to construction.

China's State Environmental Protection Administration (SEPA) used the legislation to halt 30 projects in 2004, including three hydro-power plants under the Three Gorges Project Company. Although one month later (Note as a point of reference, that the typical EIA for a major project in the USA takes one to two years.), most of the 30 halted projects resumed their construction, reportedly having passed the environmental assessment, the fact that these key projects' construction was ever suspended was notable.

A joint investigation by SEPA and the Ministry of Land and Resources in 2004 showed that 30 to 40 per cent of the mining construction projects went through the procedure of environment impact assessment as required, while in some areas only 6 to 7 per cent did so.

This partly explains why China has witnessed so many mining accidents in recent years. SEPA alone cannot guarantee the full enforcement of environmental laws and regulations, observed Professor Wang Canfa, director of the centre to help environmental victims at China University of Political Science and Law. In fact, according to Wang, the rate of China's environmental laws and regulations that are actually enforced is estimated to be barely 10 per cent.

EU

The EIA Directive on Environmental Impact Assessment of the effects of projects on the environment was first introduced in 1985 and was amended in 1997. The directive was amended again in 2003 following the 1998 signature by the EU of the Aarhus Convention on public participation in environmental matters.

The issue was enlarged to the assessment of plans and programmes by the so called SEA-Directive in 2001 which is now in force and establishes a mix of mandatory and discretionary procedures for assessing environmental impacts. Under the EU directive, an EIA must provide certain information to comply. There are seven key areas that are required:

Description of the Project

- Description of actual project and site description
- Break the project down into its key components, ie construction, operations, decommissioning
- For each component list all of the sources of environmental disturbance
- For each component all the inputs and outputs must be listed, eg, air pollution, noise, hydrology

Alternatives that have been Considered

- Examine alternatives that have been considered
- Example: in a biomass power station, will the fuel be sourced locally or nationally?

Description of the Environment

- List of all aspects of the environment that may be effected by the development
- Example: populations, fauna, flora, air, soil, water, humans, landscape, cultural heritage
- This section is best carried out with the help of local experts, eg the RSPB in the UK

Description of the Significant Effects on the Environment

- The word significant is crucial here as the definition can vary
- 'Significant' needs to be defined
- The most frequent method used here is use of the Leopold matrix
- The matrix is a tool used in the systematic examination of potential interactions
- Example: in a windfarm development a significant impact may be collisions with birds

Mitigation

- This is where EIA is most useful
- Once section 4 has been completed it will be obvious where the impacts will be greatest
- Using this information ways to avoid negative impacts should be developed
- Best working with the developer with this section as they know the project best
- Using the windfarm example again construction could be out of bird nesting seasons

Non-technical Summary

- The EIA will be in the public domain and be used in the decision making process
- It is important that the information is available to the public
- This section is a summary that does not include jargon or complicated diagrams
- It should be understood by the informed lay-person

Lack of Know-how/technical Difficulties

- This section is to advise any areas of weakness in knowledge
- It can be used to focus areas of future research
- Some developers see the EIA as a starting block for good environmental management

New Zealand

In New Zealand, EIA is usually referred to as Assessment of Environmental Effects (AEE). The first use of EIA's dates back to a Cabinet minute passed in 1974 called Environmental Protection and Enhancement Procedures. This had no legal force and only related to the activities of government departments. When the Resource Management Act was passed in 1991, an EIA was required as part of a resource consent application. Section 88 of the Act spells this out.

United States

Under United States environmental law an EIA is referred to as the Environmental Impact Statement (EIS), and originated in the National Environmental Policy Act (NEPA), enacted in the United States in 1969. Certain actions of federal agencies must be preceded by an EIS. Contrary to a widespread misconception, NEPA does not prohibit the federal government or its licensees/permittees from harming the environment, nor does it specify any penalty if the EIS turns out to be inaccurate, intentionally or otherwise. NEPA requires that plausible statements as to the prospective impacts be disclosed in advance. The purpose of NEPA process is to ensure that the decision maker is fully informed of the environmental aspects and consequences prior to making the final decision.

Usually, an agency will release a Draft Environmental Impact Statement (DEIS) for comment. Interested parties and the general public have the opportunity to comment on the draft, after which the agency will approve the Final Environmental Impact Statement (FEIS). Occasionally, the agency will later release a Supplemental Environmental Impact Statement (SEIS), especially if environmental conditions suddenly change after the issuance of the FEIS.

The adequacy of an EIS can be challenged in court. Major proposed projects have been blocked because of an agency's failure to prepare an acceptable EIS. One prominent example was the Westway landfill and highway development in and along the Hudson River in New York City.

Another prominent case involved the Sierra Club suing the Nevada Department of Transportation over its denial of Sierra Club's request to issue a supplemental EIS addressing air emissions of particulate matter and hazardous air pollutants in the case of widening US Highway 95 through Las Vegas. The case reached the 9th Circuit Court of the United States, which led to construction on the highway being halted until the court's final decision. The case was settled prior to the court's final decision.

Several US state governments that have adopted "little NEPA's," i.e., state laws imposing EIS requirements for particular state actions and some of those state laws refer to the required environmental impact studies as Environmental Impact Reports or Environmental Impact Assessments. For example, the California Environmental Quality Act (CEQA) requires an Environmental Impact Report (EIR).

These various state requirements are yielding voluminous data not just upon impacts of individual projects, but also to elucidate scientific areas that had not been sufficiently researched. For example, in a seemingly routine Environmental

Impact Report for the city of Monterey, California, information came to light that led to the official federal endangered species listing of Hickman's potentilla, a rare coastal wildflower.

ENVIRONMENTAL IMPACT ASSESSMENT OF FORESTRY PROJECTS MENT ENVIRONMENTAL IMPACT ASSESSMENT OF FORESTRY PROJECTS

The current Regulations are The Environmental Impact Assessment (Forestry) (England and Wales) Regulations 1999 [SI 1999/2228] and the Environmental Impact Assessment (Forestry) (Scotland) Regulations 1999 [SI 1999/43]. These Regulations have been further amended by

The Environmental Impact Assessment (Forestry) (England and Wales) (Amendment) Regulations 2006 and The Environmental Impact Assessment (Scotland) Amendment Regulations 2006. These Regulations came into force on 6th September 1999 and require anyone who wishes to carry out a relevant project (i.e. deforestation, afforestation, forestry roads or quarries that might have a significant effect on the environment) to obtain consent for the work from the Forestry Commission. The applicant or proposer must submit an Environmental Statement in support of the proposals to apply for consent.

You will find below some of the terms used in the EIA Regulations:

a) Appropriate Authority-The Secretary of State for Environment, Food and Rural Affairs in England, the National Assembly for Wales, and the Scottish Executive in Scotland.

b) Countryside bodies

   England-Natural England, Environment Agency. Wales-Countryside Council for Wales, Environment Agency. Scotland-Scottish Natural Heritage, Scottish Environmental Protection Agency. And any other body designated by statutory provision as having specific environmental responsibilities.

c) Determination-Taken from Regulation 15 "Determination of applications" and is the process by which we make our decision about the application for consent.

d) Forestry projects-the project work that the Forestry Commission must assess under these Regulations are deforestation (conversion to another land use), afforestation, forestry roads and forestry quarries.

e) Opinion-Our consideration of the proposals from which we will decide whether or not the project is a relevant one. If it is, the applicant must apply for consent and provide an ES.

f) Relevant project-A forestry project (afforestation, deforestation, forest roads works and forest quarry works) that is likely, by virtue of factors such as its nature, size and location, to have a significant effect on the environment and as such requires the FC's consent.

g) Scoping-A gathering of all consultees and other interested parties to discuss and agree the significant issues of concern that require to be addressed by an applicant when preparing an Environmental Statement.

h) Screening-the process by which the Forestry Commission decides whether a project "is likely to have significant effects on the environment by virtue, inter alia, of its nature, size or location". This process is a distinct one from scoping.

i) Publicity-The Forestry Commission maintains a web-based EIA Register that gives details of all the decisions we make under these Regulations. Applicants are also required to advertise details of any application for consent.

j) Thresholds-Area limits set by the Regulations below which it is not expected that the project will have a significant effect on the environment.

The Forestry Commission will assess whether:

- o the proposed project is one of the following categories-afforestation, deforestation, forest roads or quarries;
- o the area is above the relevant threshold (includes extensions to similar areas of work);
- o the project is likely to have a significant effect on the environment.

Our consent will be required to proceed with the work if your proposals meet each of these requirements. You will be asked to provide an Environmental Statement to allow us to decide whether to give consent to the project. Under the 1999 Regulations, proposals are considered to be relevant projects (i.e. to require an EIA) if they fall within the categories listed below and the work proposed is likely to have a significant effect on the environment.

a) Initial afforestation: creating new woods and forests by planting trees (on an area that has not had trees for many years). This category includes using direct seeding or natural regeneration, planting Christmas trees and short rotation coppice;

b) Deforestation: conversion of woodland to another type of land use (e.g.heathland);

c) Forest roads: constructing forestry roads, including those within a forest and those leading to a forest;

d) Forest quarries: quarrying to obtain material (rock, sand and gravel) for the formation, alteration or maintenance of forest roads.

We may serve you with an Enforcement Notice if you carry out work on a project that would have required our consent. This notice will require you to comply with the Regulations. Situations may arise where one of the above forestry projects forms part of a wider development that requires Planning Permission.

In these circumstances, any necessary EIA will usually not be dealt with under the Forestry EIA Regulations but under the parallel Town and Country Planning EIA Regulations.

Where a project has been completed since 6 September 1999, work proposals of the same type (including that on land in different ownerships) that extend it beyond the thresholds may need our consent. This also applies where the accumulated area of all adjacent existing projects plus the new proposals, exceeds the thresholds.

Making the decision Definition of relevant projects Proposals that extend a project area previously below the area thresholds

If the area of the work you are proposing falls below these thresholds then it is not likely to be a relevant project and our consent under the EIA Regulations will not be required. You may still of course require a felling licence or to get approval under the Woodland Grant Scheme for the work.

If your project is above these area thresholds, our consent for carrying out the work may be required.

You must provide the following so that we can give our opinion on whether the work you are proposing is a relevant project and requires our consent:

a) a map identifying the area and showing the extent of the project. (This should be a clear Ordnance Survey map at a scale of 1:10 000 or 1:2 500);

b) a brief description of the proposed work and its possible significant effects on the environment;

c) any other information that might be relevant, e.g. species maps, plans, photographs etc.

In the majority of instances this information will be provided as part of a Woodland Grant Scheme application (initial afforestation) or an application for a felling licence (deforestation) or Forest Plan.

If, at any time, we become aware of proposals that require our consent, we will give our opinion to the person whom we believe should have asked us. We will give our opinion, in writing, within 28 days of receiving the necessary information. However, if we need more information from you to help us decide, then we will let you have a decision within 28 days of receiving this information. If our decision will take longer we will seek to agree an appropriate period with you. If we tell you that your proposal is a relevant project, or we do not give you our opinion within the agreed timescale, you can appeal in writing to: The Scottish Ministers (in relation to Scotland); The Secretary of State for Environment, Food and Rural Affairs; The National Assembly for Wales in Wales for a "direction" as to whether the work is a relevant project. Once we have told you our decision we will publish this on an "EIA Public Register" held at Forestry Commission offices. This will show information about our opinion whether a project is a relevant project, our decision about the granting of consent and any directions that may have been made by the appropriate authority or the Scottish Ministers.

Getting our Consent

We must have the following information so that we can make a decision:-

a) a map showing the area where the project is proposed. (This should be a clear Ordnance Survey map at a scale of 1:10 000 or 1:2 500);

b) a brief description of the proposed work and the possible significant effects on the environment;

c) any other information that might be relevant e.g. species maps, plans, photographs etc.

   (You may already have supplied sufficient information for the above if you have asked for our Opinion under the EIA regulations or submitted a Woodland Grant Scheme or Forest Plan application)

d) an Environmental Statement for the work;

e) a copy of the publicity notice that you must place in newspapers.

Before you start to prepare your ES, we would encourage you to meet us and discuss the significant environmental issues that will need to be covered, or ask for information to help you.

At an early stage, it is strongly recommended that you contact the appropriate countryside bodies* and request any information they may have that is relevant to the project. [Please note that there may be a small charge for providing this information].

* In England-Natural England, English Heritage, Environment Agency; In Wales-Countryside Council for Wales, Environment Agency;

In Scotland-Scottish Natural Heritage, Scottish Environmental Protection Agency

In addition we would advise that you hold a scoping meeting with countryside bodies and other relevant organisations to help determine the issues that must be covered in the Environmental Statement. There is no set format for an ES but it must contain the standard information contained in Appendix 2. Your Statement must include everything from the list that relates to your application, placing the emphasis on the significant environmental effects. You can also get more advice about preparing an Environmental Statement in our booklet- "Preparing an Environmental Statement for a Forestry Project", only available as a PDF download from www.forestry.gov.uk.

It may help to facilitate your application if you let us have a draft copy of your ES before preparing the final copy. This will help us to decide if you have covered and addressed all the relevant issues. If we need more information to help us decide about your application, you must supply this. You can get a copy of this booklet on the Forestry Commission Grants and Licences section of the website. If you have applied for our consent and prepared an ES, you must publish a notice giving the details outlined below. We will agree at least two newspapers in which the notice must appear.

The notice must:

a) state that an application for our consent under the Regulations has been made;

b) indicate where (and at what times) copies of the application for consent can be inspected (the local Forestry Commission office or other location(s) we agree);

c) state the address where people can get copies of the application;

d) give details of any charge that may be made for copies- to cover printing, copying and distribution costs only;

e) state that comments about the application must be made to the relevant Forestry Commission Conservator within 28 days of the date of the notice.

f) You must also include the following statement: "The Forestry Commissioners may decide either to grant consent subject to the standard conditions required by Regulation 18 of the EIA Regulations, or subject to such further conditions as they see fit or refuse consent."

We will also give details of your application to the appropriate countryside body as well as the local authority with an interest in the application. They are required to give us their comments within 28 days.

We will give you our decision (and the reasons for it) in writing, once the 28-day period for comments has expired and we have considered the environmental information, any representations and any other material consideration. A notice giving details of our decision will be published in the same newspapers in which the initial notice about the application appeared. We will also inform anyone from whom we have received representations in relation to the application.

When granting consent, we will specify the date by which you must start and finish the work. The start date will be no later than five years from the date of the consent and the finish date will be no later than ten years from the date of the consent. We may also apply other conditions to the consent that we deem necessary to ensure that the environment is protected from the impact of the project. These may be in addition to the normal Grant Scheme or Felling Licence conditions. Publicity about the application Consultation about the application Giving our decision about your application Standard consent conditions.

If you have applied for consent, you can appeal about the decision we make to:

*The Scottish Ministers (in relation to Scotland);

*The Secretary of State for Environment, Food and Rural Affairs in England;

*The National Assembly for Wales in Wales;

You must do this within 28 calendar days of receiving notice of our decision or within any other period that the Government authority* may agree.

You may only appeal if the Forestry Commissioners have:

a) refused the application;

or

b) have granted consent with conditions (additional to the standard conditions); or

c) we have asked you to start and finish the work within a period that is less than the maximum period permitted by the Regulations. In addition, any person who feels aggrieved by the granting of consent can make an application to the Court if they feel that their interests are prejudiced. They must do this within 6 weeks of consent being granted.

If we discover that you have carried out work that would have required our consent, or have breached the conditions of consent, we may serve you with an Enforcement Notice. This will require you to do one or more of the following:

a) stop the work;

b) apply for consent;

c) restore the land to its condition before the work was started;

d) carry out work to secure compliance with the conditions of the consent;

e) remove or alleviate any damage to the environment that has been caused by the work.

You can appeal against the serving of the Enforcement Notice to:

*The Scottish Ministers (in relation to Scotland);

*The Secretary of State for Environment, Food and Rural Affairs in England;

*The National Assembly for Wales in Wales;

You must do this within 28 calendar days of receiving notice of our decision or within any other period that the Government authority* may agree.

If you require more information or help, including how to make appeals, contact your local Forestry Commission office or: Grants and Licences, Forestry Commission Headquarters, 231 Corstorphine Road, Edinburgh EH12 7AT

Appealing against our decision Carrying out work without our consent or breaching the conditions of consent Appealing against the enforcement notice Getting more information.

Definition of "sensitive areas":

For the purposes of the the definitions of "sensitive areas" are as follows:

a) National Nature Reserve or Site of Special Scientific Interest;

b) National Parks;

c) The Broads;

d) A World Heritage Site;

e) Scheduled Ancient Monuments;

f) Area of Outstanding Natural Beauty;

g) Natural Heritage or National Scenic Areas;

h) A site designated or identified as a candidate Special Area of Conservation;

i) The New Forest Heritage Area;

j) A site classified or proposed as a Special Protection Area.

Information for inclusion in environmental statements Part 1.

1) Description of the project, including in particular:

   a) a description of the physical characteristics of the whole project and the land-use requirements during the construction and operational phases;

   b) a description of the main characteristics of the production processes, for instance, nature and quantity of the materials used;

   c) an estimate, by type and quantity, of expected residues and emissions (water, air and soil pollution, noise, vibration, light heat, radiation, etc) resulting from the operation of the proposed project.

2) An outline of the main alternatives studied by the developer and an indication of the main reasons for the developer's choice, taking into account environmental effects.

3) A description of the aspects of the environment likely to be significantly affected by the proposed project, including, in particular, human beings, fauna, flora, soil, water, air, climatic factors, material assets including the architectural and archaeological heritage, landscape and the inter-relationship between the above factors.

4) A description of the likely significant effects of the proposed project on the environment, which should cover the direct effects and any indirect, secondary, cumulative, short, medium and long-term, permanent and temporary, positive and negative effects of the project, resulting from:

   a) the existence of the project;

   b) the use of natural resources;

   c) the emission of pollutants, the creation of nuisances, and the elimination of waste, and the description by the developer of the forecasting methods used to assess the effects on the environment.

5) A description of the measures proposed to prevent, reduce and, where possible offset, any significant adverse effects on the environment.

6) A non-technical summary of the information provided under paragraphs 1 to 5 above.

7) An indication of any difficultries (technical deficiencies or lack of knowledge) encountered by the developer in compiling the required information.

GUIDE TO PROCEDURES

Environmental impact assessment (EIA) is an important procedure for ensuring that the likely effects of new development on the environment are fully understood and taken into account before the development is allowed to go ahead. This booklet, which is intended primarily for developers and their advisers, explains how European Community (EC) requirements for the environmental impact assessment of major projects have been incorporated into consent procedures in the UK. It revises the booklet 'Environmental Assessment: A Guide to the Procedures', first published in 1989, to take account of the requirements of Directive 97/11/EC, which was adopted on 3 March 1997 and came into effect on 14 March 1999. Directive 97/11/EC amends the original Directive 85/337/EEC on 'The assessment of the effects of certain public and private projects on the environment', which came into effect in July 1988. The text of the Directive as amended (with amendments shown in italics) is reproduced in Appendix 1 to this booklet. Throughout this booklet, references to `the Directive' mean Directive 85/337/EEC as amended by Directive 97/11/EC.

The effect of the Directive is to require environmental impact assessment to be carried out, before development consent is granted, for certain types of major project which are judged likely to have significant environmental effects. Parts 1 and 2 of this booklet explain the procedures which apply to projects falling within the scope of the Directive and requiring planning permission in England and Wales.

For such projects the Directive was given legal effect through the Town and Country Planning which came into force on 14 March 1999 and apply to relevant planning applications lodged on or after that date. The full text of the Regulations is published separately by The Stationery Office. For ease of reference, those parts of the Regulations which list the types of project to which they apply, and specify what information an environmental statement must contain, are reproduced in Appendices 2 to 4 to this booklet. The booklet is not intended to be an authoritative interpretation of the law and does not remove the need to refer to the Regulations.

Formal guidance on procedures under the Regulations, directed principally at local planning authorities, was issued in DETR Circular 2/99 (Welsh Office Circular 11/99). Although the present booklet, like its predecessor, is meant to be reasonably self-contained, developers may need to refer to that Circular, particularly for fuller information on how planning authorities are expected to judge the significance of a project's likely effects for the purpose of deciding whether environmental impact assessment is required. Parts 1 and 2 also give some general guidance, applicable to all types of project, on the nature of environmental impact assessment and on the practical aspects of preparing an environmental statement. Part 3 gives a brief account of the procedures which apply to other projects within the scope of the Directive but which are not approved under planning procedures, for example, motorways, harbour works and long distance pipe-lines. It also deals very briefly with environmental impact assessment procedures in Scotland and Northern Ireland. For the detailed requirements, reference will need to be made to the relevant statutory instruments and associated guidance.

Throughout this booklet, the term 'environmental impact assessment' (EIA) is used to describe the whole process whereby information about the environmental effects of a project is collected, assessed and taken into account in reaching a decision on whether the project should go ahead or not. The process was formerly referred to in the UK as 'environmental assessment' (EA). An 'environmental statement' is a document setting out the developer's own assessment of a project's likely environmental effects, which is prepared and submitted by the developer in conjunction with the application for consent. Go to table of contents

ENVIRONMENTAL IMPACT ASSESSMENT AND PROJECTS WHICH REQUIRE PLANNING PERMISSION

The term 'environmental impact assessment' (EIA) describes a procedure that must be followed for certain types of project before they can be given 'development consent'. The procedure is a means of drawing together, in a systematic way, an assessment of a project's likely significant environmental effects. This helps to ensure that the importance of the predicted effects, and the scope for reducing them, are properly understood by the public and the relevant competent authority before it makes its decision.

Environmental impact assessment enables environmental factors to be given due weight, along with economic or social factors, when planning applications are being considered. It helps to promote a sustainable pattern of physical development and land and property use in cities, towns and the countryside. If properly carried out, it benefits all those involved in the planning process.

From the developer's point of view, the preparation of an environmental statement in parallel with project design provides a useful framework within which environmental considerations and design development can interact. Environmental analysis may indicate ways in which the project can be modified to avoid possible adverse effects, for example through considering more environmentally friendly alternatives. Taking these steps is likely to make the formal planning approval stages run more smoothly.

For the planning authority and other public bodies with environmental responsibilities, environmental impact assessment provides a basis for better decision making. More thorough analysis of the implications of a new project before a planning application is made, and the provision of more comprehensive information with the application, should enable authorities to make swifter decisions. While the responsibility for compiling the environmental statement rests with the developer, it is expected that the developer will consult those with relevant information, and the Regulations specifically require that public authorities which have information in their possession which is relevant to the preparation of the environmental statement should make it available to the developer.

The general public's interest in a major project is often expressed as concern about the possibility of unknown or unforeseen effects. By providing a full analysis of a project's effects, an environmental statement can help to allay fears created by lack of information. At the same time, early engagement with the public when plans are still fluid can enable developers to make adjustments which will help to secure a smoother passage for the proposed development and result in a better environmental outcome. The environmental statement can also help to inform the public on the substantive issues which the local planning authority will have to consider in reaching a decision. It is a requirement of the Regulations that the environmental statement must include a description of the project and its likely effects together with a summary in non-technical language. One of the aims of a good environmental statement should be to enable readers to understand for themselves how its conclusions have been reached, and to form their own judgements on the significance of the environmental issues raised by the project.

Environmental impact assessment can therefore be helpful to all those concerned with major projects. The following paragraphs describe the procedures for deciding whether EIA is necessary in a particular case and, where it is, for carrying out the assessment. The procedure is intended to make the most of the potential benefits of EIA, while keeping the process as simple and flexible as possible, and avoiding any duplication of existing planning procedures.

WHEN IS ENVIRONMENTAL IMPACT ASSESSMENT NEEDED?

The Regulations apply to two separate lists of projects:

i. 'Schedule 1 projects', for which EIA is required in every case;

ii. 'Schedule 2 projects', for which EIA is required only if the particular project in question is judged likely to give rise to significant environmental effects. Lists of Schedule 1 and Schedule 2 projects are given in, respectively, Appendices 2 and 3 to this booklet.

For Schedule 1 projects, whether or not a particular project falls within the scope of the Regulations will normally be clear: several of the definitions of Schedule 1 projects incorporate an indication of scale, in the form of a quantified threshold, which clearly identifies the projects requiring EIA. Where there is any doubt about a project's inclusion in Schedule 1, the procedures described in below can be used to obtain an opinion from the planning authority or a direction from the Secretary of State (or, in Wales, the National Assembly for Wales).

For the much longer list of Schedule 2 projects, the issue turns on the likelihood of 'significant environmental effects'. For the different types of project described in column 1 of Schedule 2, the 1999Regulations introduced a system of thresholds and criteria, shown in column 2,as a method of discounting development which is not likely to have significant effects on the environment. For development where the applicable threshold or criterion is not exceeded or met, EIA is not normally required. However, even where the threshold or criterion is not met or exceeded, EIA may be required if the proposed development is in, or partly in, a 'sensitive area.' In exceptional circumstances the Secretary of State (or, in Wales, the National Assembly for Wales) may exercise his power under the Regulations to direct that a particular type of Schedule 2 development requires EIA even if it is not to be located in a sensitive area and does not exceed or meet the applicable threshold or criterion.

The more environmentally sensitive the location, the more likely it is that the effects of development will be significant and that EIA will be required. That is why the thresholds and criteria do not apply where development is proposed in, or partly in, a `sensitive area' as defined in the Regulations. Such areas include Sites of Special Scientific Interest (SSSIs), National Parks, Areas of Outstanding Natural Beauty, the Broads, World

Heritage Sites and scheduled monuments. There is no general presumption that every Schedule 2 development in a sensitive area will require EIA. Nevertheless, in the case of development to be located in or close to SSSIs, especially those which are also international conservation sites such as Ramsar sites or Special Protection Areas for birds, the likely environmental effects will often be such as to require EIA.

HOW 'SIGNIFICANCE' WILL BE ASSESSED

Developments which meet or exceed the applicable threshold are considered on a caseby-case basis. For the purpose of determining whether EIA is necessary, those of the selection criteria set out in Schedule 3 to the Regulations which are relevant to the proposed development, must be taken into account. The selection criteria fall into the three broad headings: characteristics of the development, location of the development, and characteristics of the potential impact.

For obvious reasons there can be no general definition of what constitutes significance.

General guidance on how to assess `significance' is contained in DETR Circular 2/99 (Welsh Office Circular 11/99);and rulings may be obtained from the local planning authority or the Secretary of State (or, in Wales, the National Assembly for Wales) on whether EIA is required in particular cases. Essentially the Circular suggests that there are three main criteria of significance:

i. major developments which are of more than local importance;

ii. developments which are proposed for particularly environmentally sensitive or vulnerable locations;

iii. developments with unusually complex and potentially hazardous environmental effects.

These are very general guidelines and, to assist in their application to particular cases, the Circular also sets out indicative thresholds and criteria by reference to particular categories of development listed in Schedule 2 to the Regulations.

These are reproduced in the last column of the Table in Appendix 3 to this booklet. It will be obvious that none of these guidelines can be applied as hard and fast rules; circumstances are bound to vary greatly from case to case. Some large-scale projects which exceed the indicative thresholds may not be significant enough to require EIA; some smaller projects, particularly in sensitive locations, may be candidates for EIA. Nevertheless, the guidance in the Circular should provide a starting point for consideration by the developer and the planning authority of the need for EIA. If the matter is referred to the Secretary of State (or, in Wales, the National Assembly for Wales), he will have regard to the published criteria.

OBTAINING A RULING ON THE NEED FOR EIA

Where there is a possibility that a proposed development will require environmental impact assessment, developers are advised to consult the relevant planning authority well in advance of a planning application. Developers can decide for themselves that a given project falls within the scope of the Regulations so that an environmental statement will be needed. But the Regulations also provide a procedure which enables developers to apply to the planning authority for an opinion ('screening opinion') on whether EIA is needed in a particular case, as soon as a basic minimum of information can be provided about the proposal. This must include a plan on which the site of the proposed development is identified, and a brief description of its nature and purpose and of its possible effects on the environment. This can, of course, be supplemented with other information if the developer wishes.

Where such information can be provided, the developer may approach the planning authority at any time for an opinion on the need for EIA. This can be done well in advance of any formal planning application, although any approach to the planning authority before the planning application stage is entirely voluntary. Where such an approach is made, the planning authority must give its opinion within three weeks, unless the developer agrees to a longer period. The planning authority may request further information from the developer, but this in itself does not extend the three-week time limit, unless the developer agrees. The planning authority must make its determination available for public inspection at the place where the appropriate register (or relevant section of that register) is kept.

Where the planning authority expresses the opinion that a particular proposal requires EIA-whether in response to a request from a developer prior to a planning application, or following a planning application-it must provide a written statement giving clear and precise reasons for its opinion. Both that statement and the developer's application for an opinion are then made available for public inspection at the same place as the register.

A developer who is dissatisfied with the planning authority's opinion that EIA is required may refer the matter to the Secretary of State (or, in Wales, the National Assembly for Wales). The developer is simply required to copy the relevant papers to the Government Office in the region concerned (or the Assembly) and add any representations which are considered to be appropriate in the light of the planning authority's statement.

The Secretary of State (or the Assembly) will then normally give a direction within three weeks of the developer's application; and, if the direction is to the effect that EIA is required, it will be accompanied by a statement of reasons. The developer may, when requesting a screening opinion from the planning authority, simultaneously request an opinion on what should be included in any environmental statement.

The broad intention of this procedure is to ensure that developers can obtain a clear ruling on the need for EIA well before they reach the stage of lodging a formal planning application. This should minimise the possibility of delay or uncertainty at that stage. Where the matter is not raised until a formal planning application is lodged, the developer risks serious delay if either the planning authority or the Secretary of State (or, in Wales, the National Assembly for Wales) rules that an environmental statement must be prepared. No action will be taken on the planning application until the developer has prepared an environmental statement and submitted it to the planning authority.

In most cases this procedure will give developers a firm decision on the need for EIA as soon as they can provide basic information about their project. There may occasionally be cases where the receipt of information directly leads the Secretary of State (or, in Wales, the National Assembly for Wales) to issue a direction when there has been no request for one from the developer; or to overrule a planning authority's opinion that EIA is not required; or even, very exceptionally, to reverse an earlier direction. This could, for instance, happen where a development proposal came to the Secretary of State's (or the Assembly's) notice on 'call-in'; or where third party representations drew attention to aspects of a proposed development which were not known to the Secretary of State (or the Assembly) when the initial direction was given. Exceptionally, also, a planning authority which has given a pre-application opinion that EIA is not required might consider it necessary to reverse that decision when the planning application is formally submitted. In that case the developer could, of course, apply to the Secretary of State (or the Assembly) for a direction.

'PERMITTED DEVELOPMENT RIGHTS' (PDRS)

Developments which do not require planning permission because of the provisions of the Town and Country Planning (General Permitted Development) PDRs provided that they do not fall into Schedule 1 or 2 of the EIA Regulations. For developments that do fall within Schedule 1 or 2, the general position is as follows. Schedule 1projects are not permitted development, and always require the submission of a planning application and an environmental statement. PDRs for Schedule 2 projects which either exceed or meet the applicable threshold or criterion, or are wholly or partly in a sensitive area, are also withdrawn, unless the local planning authority has adopted a screening opinion (or the Secretary of State (or, in Wales, the National Assembly for Wales) has directed) to the effect that

EIA is not required. There are exceptions to these provisions in the case of the following classes in Schedule 2 to the 1995 Order: Part 7, Class D of Part 8, Part 11, Class B of Part 12, Class F (a) of Part 17, Class A of Part 20, Class B of Part 20, and Class B of Part

These exceptions exist for a variety of reasons. For example, some relate to projects subject to alternative consent procedures, and others to projects begun before Directive 85/337/EEC came into operation. Further information about the exceptions, and about permitted development and EIA generally, can be found in DETR Circular 2/99 (Welsh Office Circular 11/99).

SIMPLIFIED PLANNING ZONES AND ENTERPRISE ZONES

Special considerations apply to projects proposed for simplified planning zones (SPZs) and enterprise zones (EZs).

Simplified Planning Zones

All Schedule 1 projects are excluded from the scope of SPZs and therefore require EIA as part of the application for planning permission. Where the terms of SPZ schemes would permit Schedule 2 projects to be undertaken without specific planning permission, developers must notify the planning authority that they intend to undertake such development. This notification will give the planning authority and, where appropriate, the Secretary of State (or, in Wales, the National Assembly for Wales) the opportunity to consider whether the project is likely to give rise to significant environmental effects. If so, planning permission will be required in the normal way and an environmental statement must be prepared. If not, the project will enjoy the benefit of the general permission granted by the SPZ scheme, and no separate application for planning permission will be necessary.

Enterprise Zones

If the planning scheme allows for Schedule 1 projects, EIA must be carried out. As regards Schedule 2 projects, the same procedures apply as those for SPZs described above.

The arrangements in SPZs and EZs are explained in DETR Circular 2/99 (Welsh Office Circular 11/99).

Go to table of contents

Preparing an Environmental Statement: the Planning Procedures

- Preliminary consultations
- Content of the environmental statement
- Statutory and other consultees; the general public
- Techniques of assessment; sources of advice
- Submission of environmental statement with planning application
- Handling by the planning authority
- Requests for further information
- Determination of application
- Appeals and call-ins
- Procedural stages

The developer is responsible for preparing the environmental statement which must be submitted with the planning application. A developer may choose to engage consultants for some or all of the work. The preparation of the statement should be a collaborative exercise involving discussions with the local planning authority, statutory consultees and possibly other bodies as well. There is no prescribed form of statement, provided that the requirements of the Regulations are met. The Regulations enable a developer, before making a planning application, to ask the local planning authority for its formal opinion ('scoping opinion') on the information to be included in an environmental statement. The developer can therefore be clear as to what the authority considers are the aspects of the environment which would be affected. The request may be made at the same time as for the screening opinion, and must be accompanied by the same information as for the screening opinion. A developer may also wish to submit a draft outline of the environmental statement, indicating what the developer thinks the main issues are, as a focus for the authority's consideration. The authority must consult certain bodies and the developer before adopting a scoping opinion, and must adopt the scoping opinion within five weeks of receiving the request. This period may be extended provided that the developer agrees. The scoping opinion must be kept available for public inspection for two years, with the request (including documents submitted by the developer as part of that request), at the place where the planning register is kept.

There is no provision for an appeal to the Secretary of State (or, in Wales, the National Assembly for Wales) if the developer and local planning authority disagree about the content of an environmental statement. If an authority fails to adopt a scoping opinion within five weeks (or any agreed extension), the developer may apply to the Secretary of State (or the Assembly) for a scoping direction. The application must be accompanied by the previous documents relating to the request for a scoping opinion, plus any additional representations the developer wishes to make. The developer should also send a copy of the request and any representations to the authority. The Secretary of State (or the Assembly) must make a scoping direction within five weeks of receiving the request, or such longer period as he may reasonably require. He must consult the developer and certain bodies beforehand. A copy of the scoping direction will be sent to the developer and the authority.

The aim should be to provide as systematic and objective an account as is possible of the significant environmental effects to which the project is likely to give rise. Where the statement embodies or summarises the conclusions of more detailed work, sufficient information should be provided to enable those who wish to do so to verify the statement's conclusions and to identify the source of the information provided. The environmental statement must contain a non-technical summary which will enable non-experts to understand its findings.

Preliminary Consultations

One of the main emphases of the process of environmental impact assessment is on the need for fuller and earlier consultation by the developer with bodies which have an interest in the likely environmental effects of the development proposal. If important issues are not considered at a very early stage, they may well emerge when a project's design is well advanced, and necessitate rethinking and delay. Ideally, EIA should start at the stage of site selection and (where relevant) process selection, so that the environmental merits of practicable alternatives can be properly considered.

While a developer is under no formal obligation to consult about the proposal before the submission of a formal planning application, there are good practical reasons for doing so. Authorities will often possess useful local and specialised information which is relevant to a project's design, and officers may be able to give preliminary advice about local problems and about those aspects of the proposal that are likely to be of particular concern to the authority.

The timing of such informal consultations is at the developer's discretion; but it will generally be advantageous for them to take place as soon as the developer is in a position to provide sufficient information about the proposal to form a basis for discussion. The developer can ask that any information provided at this preliminary stage should be treated in confidence by the planning authority and any other consultees. If the developer is seeking a formal opinion from the planning authority on the need for environmental impact assessment, the information about the project which accompanies that request will be made public by the authority.

Content of the Environmental Statement

Developers and authorities should discuss the scope of an environmental statement before its preparation is begun. The formal requirements as to the content of environmental statements are set out in Schedule 4 to the Regulations, which is reproduced in Appendix 4 to this booklet. The information is given under separate headings: Part I and Part II. The statement must include at least the information included in Part II, and such of the information in Part I as is reasonably required to assess the environmental effects of the development and which the applicant can reasonably be required to complete. As a practical guide to the range of issues which may need to be considered, developers may find it helpful to use the checklist at Appendix 5 to this booklet as a basis for their discussions with the planning authority. The checklist is not meant to be regarded as a prescribed framework for all environmental statements. Its main purpose is to act as a guide or agenda for the preliminary discussions about the scope of the statement.

The comprehensive nature of the checklist at Appendix 5 should not be taken to imply that all environmental statements should cover every conceivable aspect of a project's potential environmental effects at the same level of detail. They should be tailored to the nature of the project and its likely effects. Whilst every environmental statement should provide a full factual description of the project, the emphasis of Schedule 4 is on the main or significant effects to which a project is likely to give rise.

In some cases, only a few of the aspects set out in the checklist will be significant in this sense and will need to be discussed in the statement in any great depth. Other issues may be of little or no significance for the particular project in question, and will need only very brief treatment, to indicate that their possible relevance has been considered.

It should be noted that developers are now required to include in the environmental statement an outline of the main alternative approaches to the proposed development that they may have considered, and the main reasons for their choice. It is widely regarded as good practice to consider alternatives, as it results in a more robust application for planning permission. Also, the nature of certain developments and their location may make the consideration of alternatives a material consideration. Where alternatives are considered, the main ones must be outlined in the environmental statement.

Even where a local planning authority has adopted a scoping opinion, the developer is responsible for the content of the statement which is finally submitted. Developers should bear in mind that planning authorities have powers to call for additional information when considering environmental statements and planning applications, and that they are likely to use those powers if they consider that aspects of a submitted environmental statement are inadequate. There is no provision for any disagreement between the developer and the planning authority over the content of an environmental statement to be referred to the Secretary of State (or, in Wales, the National Assembly for Wales), except through normal planning appeal procedures.

Developers should consider at an early stage whether an assessment of environmental effects may also be required under another European Community Directive, such as the Habitats Directive (92/43/EEC), the Wild Birds Directive(79/409/EEC), the Integrated Pollution Prevention and Control Directive(96/61/EC) or the Control of Major Accident Hazards Directive (96/82/EC).Although the requirements of these and of the EIA Directive are all independent of each other, there are clearly links between them. Where more than one regime applies, developers could save unnecessary time and effort if they identify and co-ordinate the different assessments required. Advice on the links between the EIA system and the requirements of the Habitats Regulations is offered in PPG 9on Nature Conservation (or, in Wales, Planning Guidance (Wales) Planning Policy First Revision), and on the links between the Town and Country Planning system and the IPPC authorisation system in PPG 23 on Planning and Pollution Control (or, in Wales, Planning Guidance (Wales) Planning Policy First Revision and Planning Guidance (Wales) Technical Advice Note (Wales) 5 'Nature Conservation and Planning').

STATUTORY AND OTHER CONSULTEES; THE GENERAL PUBLIC

The Regulations give a particular role in environmental impact assessment to those public bodies with statutory environmental responsibilities who must be consulted by the planning authority before a Schedule I or a Schedule 2 planning application is determined. A full list of these statutory consultation bodies is given in Appendix 6 to this booklet.

Where the planning authority or the Secretary of State(or, in Wales, the National Assembly for Wales) rules that EIA is required, those bodies which are statutory consultees for the particular project in question will be notified and the developer will be informed accordingly.

The effect of this notification is to put those bodies under an obligation to provide the developer (on request) with any information in their possession which is likely to be relevant to the preparation of the environmental statement. An example might be information held by English Nature (or, in Wales, the Countryside Council for Wales) about the ecology of a particular area, which could be relevant to the assessment of a project's likely effects.

It is up to the developer to approach the statutory consultees and indicate what sort of information would be helpful in preparing the environmental statement. The obligation on statutory consultees relates only to information already in their possession; they are not required to undertake research on behalf of the developer. Nor, at this stage, would consultees be expected to express a view about the merits of the proposal; their views on merits are invited at a later stage. Consultees may make a reasonable charge to cover the cost of making information requested by a developer available.

Developers should also consider whether to consult the general public, and non-statutory bodies concerned with environmental issues, during the preparation of the environmental statement. Bodies of this kind may have particular knowledge and expertise to offer. Some are national organisations, for instance, the Royal Society for the Protection of Birds; inmost areas there are also active local amenity societies and environmental groups.

While developers are under no obligation to publicise their proposals before submitting a planning application, consultation with local amenity groups and with the general public can be useful in identifying key environmental issues, and may put the developer in a better position to modify the project in ways which would mitigate adverse effects and recognise local environmental concerns. It will also give the developer an early indication of the issues which are likely to be important at the formal application stage if, for instance, the proposal goes to public inquiry.

TECHNIQUES OF ASSESSMENT; SOURCES OF ADVICE

Extensive literature is available on how to assess the effects on the environment of particular processes and activities. The assessment techniques used, and the degree of detail in which any particular subject is treated in an environmental statement, will depend on the character of the proposal, the environment which it is likely to affect, and the information available. While a careful study of the proposed location will generally be needed (including environmental survey information), original scientific research will not normally be necessary. The local planning authority and statutory consultees may be able to advise the developer on sources of specialist information, for example, about particular local conditions.

Environmental statements will often need to recognise that there is some uncertainty attached to the prediction of environmental effects. Where there is uncertainty, it needs to be explicitly recognised. Uncertainty is not in itself a reason for discounting the importance of particular potential environmental effects, simply because other effects can be more confidently predicted.

Submission of Environmental Statement with Planning Application

To enable a planning application to be processed as quickly as possible, it is in the developer's interest to submit an environmental statement at the same time as the application is made. It will be for the planning authority to judge how much information is required in the particular case, but the preparation of an environmental statement is bound to require the developer to work out proposals in some detail; otherwise any thorough appraisal of likely effects will be impossible. Where an application is in outline, the planning authority will still need to have sufficient information on a project's likely effects to enable it to judge whether the development should take place or not. The information given in the environmental statement will have an important bearing on whether matters may be reserved in an outline permission; it will be important to ensure that the development does not take place in a form which would lead to significantly different effects from those considered at the planning application stage.

When the developer submits an environmental statement at the same time as the planning application, three further copies must also be submitted for onward transmission by the planning authority to the Secretary of State (or, in Wales, the National Assembly for Wales). The developer is also required to provide the planning authority with sufficient copies of the environmental statement to enable one to be sent to each of the statutory consultees. Alternatively, the developer may send copies of the statement directly to the consultees. When submitting the application, the developer must inform the planning authority of the name of every body-whether or not it is a statutory consultee-to which a copy of the statement has been sent.

The developer should make a reasonable number of copies of the statement available for members of the public. A reasonable charge reflecting printing and distribution costs may be made.

HANDLING BY THE PLANNING AUTHORITY

Where the environmental statement is submitted with the planning application, the local planning authority will arrange for a notice to be published in a local newspaper and displayed at or near the site of the proposed development. If the environmental statement is submitted after the planning application, responsibility for publicising it falls to the developer.

The planning authority will place the planning application on Part I of the planning register, together with the environmental statement. The authority and the developer may wish to consider the need for further publicity at this stage, for example, publication of further details of the project in a local newspaper, or an exhibition. The planning authority will also need to notify statutory consultees of the application (unless the developer has already done so) and invite them to comment on the environmental statement. Consultees must be allowed at least 14 days from receipt of the statement in which to comment before a decision is taken. It will often be useful for the planning authority to discuss the project with consultees who have a particular interest in its environmental effects before reaching its conclusions on the planning applications.

The copies of the environmental statement that are forwarded by the planning authority to the Secretary of State (or, in Wales, the National Assembly for Wales), will assist in monitoring. In those exceptional circumstances where a proposed development is likely to have significant effects on the environment in another country, they will also enable certain international obligations for exchange of information with other countries to be met.

Requests for Further Information

Where the planning authority considers that the information provided in the developer's environmental statement, together with that available to the authority from other sources, is insufficient to permit a proper evaluation of the project's likely environmental effects, the authority can require further information, or evidence to verify the information that has already been provided.

The use of these powers should not normally be necessary, especially if the parties have worked together during the preparation of the environmental statement. Nevertheless, further consultation between the planning authority and the developer may be necessary at this stage, in particular to consider comments made by consultees and, possibly, amendments to the proposal to meet objections that have been raised. Where an authority considers that it does not have the necessary expertise to evaluate the information contained in an environmental statement, it may decide to seek advice from consultants or other suitably qualified persons or organisations.

Determination of Application

The planning authority is required to determine a planning application which is the subject of environmental impact assessment within 16 weeks from the date of receipt of the environmental statement, unless the developer agrees to a longer period. In determining the application, the authority is, of course, required to have regard to the environmental statement, as well as to other material considerations. As with any other planning application, the planning authority may refuse permission or grant it with or without conditions.

The planning authority cannot take the view that a planning application is invalid because it considers that an inadequate environmental statement has been submitted or because the developer has failed to provide any further information required under the powers described above. However, if the developer fails to provide enough information to complete the environmental statement, the application can be determined only by refusal.

Appeals and Call-ins

The right of appeal to the Secretary of State (or, in Wales, the National Assembly for Wales) against an adverse decision by a planning authority (or against an authority's failure to determine an application within the 16-week time limit) is the same for planning applications to which the EIA Regulations apply as for other applications. Similarly, the Secretary of State's (or the Assembly's) power to call in a planning application applies in these cases. Where an environmental statement has been prepared to accompany a planning application, the information which it contains will be among the material considerations which an Inspector will take into account in considering an appeal. The Secretary of State (or the Assembly) and Inspectors, like the planning authority, have power to request the developer to provide further information where they consider that the environmental statement is inadequate as it stands. Any additional information provided by the developer in response to such a request will be made available to all parties to an appeal.

Procedural Stages

Appendix 7 to this booklet provides illustrative flowcharts for the five main procedural stages: application to the planning authority for a screening opinion; application to the Secretary of State (or, in Wales, the National Assembly for Wales) for a screening direction; application to the planning authority for a scoping opinion; application to the Secretary of State (or the Assembly) for a scoping direction; and submission of an environmental statement to the planning authority in conjunction with a planning application.

Arrangements for other Projects

- The trunk road network
- Oil and gas pipe-lines
- Offshore oil and gas projects
- Nuclear power stations
- Other power stations and overhead power lines
- Forestry projects
- Land drainage improvements
- Ports and harbours
- Marine fish farming
- Marine dredging for minerals
- Transport and Works Act 1992
- Projects arising in Scotland and Northern Ireland

However, those Parts specifically describe the arrangements for EIA of projects in England and Wales which are approved through the planning system. Part 3 gives brief guidance on the legislative provisions on EIA for projects insofar as they are outside the scope of the planning system.

For all the following projects the Regulations described, which implement the Directive, include provisions for seeking a scoping opinion from the competent authority, for consulting statutory and other bodies, for the submission of an environmental statement in the prescribed form, and for publicising the statement. Where applicable, the Regulations disapply any thresholds for Annex II projects which are at least partly in a 'sensitive area', this term being defined in the same or almost the same way as in the planning EIA Regulations.

PROJECTS NOT SUBJECT TO PLANNING CONTROL

The Trunk Road Network

EIA is mandatory for projects to construct new motorways and certain other roads, including those with four or more lanes, and for certain road improvements.

Roads in England and Wales for which the Secretary of State for Transport or the National Assembly for Wales is the highway authority, are approved under procedures set out in the Highways Act 1980. These roads-the trunk road network-currently include nearly all motorways. The Department concerned will normally consult the public widely about alternative routes before selecting a preferred route for a new road. Once the preferred route has been announced, detailed design work is carried out leading to the publication of statutory orders. In most cases these orders are the subject of a public inquiry held by an independent inspector.

For these roads, the Directive has been implemented in England and Wales by the Highways (Assessment of Environmental Effects) Regulations 1999 (SI No 369) http://www.legislation.hmso.gov.uk/si/si1999/19990369.htm. These Regulations replace the Highways (Assessment of Environmental Effects) Regulations 1988 (SI No 1241), except for projects begun before the new Regulations came into force on 13 March 1999.

The Regulations incorporate an exclusive threshold for the purpose of screening Annex II projects: only those in which the area of the completed works and the area occupied during the road construction or improvement period exceeds one hectare, or are wholly or partly in a sensitive area, need to be screened for EIA.

Roads developed by local planning authorities, and roads developed by private developers, require planning permission and the provisions described in Parts I and 2 of this booklet apply to such roads.

Oil and Gas Pipe-lines

Oil and gas pipe-lines with a diameter of more than 800 millimetres and a length of more than 40 kilometres fall within Annex I of the Directive. Pipe-line installations which are below either of these thresholds fall within Annex II. Oil and gas pipe-lines 10 miles long or less are approved under planning legislation.

Onshore Pipe-line Works

Proposed onshore pipes (but not those of public gas transporters-see below) which are more than 10 miles long require a pipe-line construction authorisation (PCA) from the Secretary of State for Trade and Industry under the Pipe-lines Act 1962. The Pipe-line Works (Environmental Impact Assessment) Regulations 2000 (SI No 1928), which replace the provisions relating to pipe-lines in the Electricity and Pipe-line Works (Assessment of Environmental Effects) Regulations 1990, implement the Directive in England, Wales and Scotland in respect of pipe-lines which require a PCA. They require 'relevant pipe-line works' to be subject to EIA unless the Secretary of State makes a determination that EIA is not necessary. Public gas transporter pipe-line works

Under the licensing regime introduced by the Gas Act 1995, companies wishing to convey gas may be licensed as public gas transporters. As such, they are exempt from the need to obtain authorisation for their pipes under the Pipe-lines Act 1962.

The Public Gas Transporter Pipe-line Works (Environmental Impact Assessment) Regulations 1999 came into force on 15 July 1999 and implement the Directive in England, Wales and Scotland. They require a public gas transporter proposing to undertake pipe-line works which fall in Annex I of the Directive to submit an environmental statement and apply to the Secretary of State for Trade and Industry for consent to carry them out. Pipe-line works in Annex II of the Directive may be subject to EIA if they have a design operating pressure exceeding

7 bar gauge or either wholly or in part cross a sensitive area. In these circumstances, the public gas transporter must, before commencing construction, either obtain a determination from the Secretary of State that an environmental statement is not required, or give notice that it intends in any event to produce an environmental statement. The Regulations also provide for the Secretary of State to require an environmental statement where proposed works do not meet these criteria but nevertheless it is considered that there are likely to be significant environmental effects.

Offshore Oil and Gas Projects

The Directive covers the extraction of petroleum and natural gas for commercial purposes where the amount extracted exceeds 500 tonnes per day in the case of petroleum and 500 000 cubic metres per day in the case of gas.

For such projects, the Offshore Petroleum Production and Pipe-lines (Assessment of Environmental Effects) Regulations 1999 (SI No 360) implement the Directive. They apply to the whole of the UK. They replace the Offshore Petroleum and Pipe-lines (Assessment of Environmental Effects) Regulations 1998 (SI No 968), which are revoked except for saving provisions for applications for consent received before 14 March 1999.

The Regulations are linked to licences granted by the Secretary of State for Trade and Industry under the Petroleum Act 1998. Such licences require the Secretary of State's consent to be obtained to the drilling of a well, the getting of petroleum (where the amount exceeds 500 tonnes per day in the case of oil and 500 000 cubic metres per day in the case of gas) and the erection of any structure in connection with a development. They require, subject to an exception for floating installations whose use commenced before 30 April 1998 (when the 1998 Regulations came into force), the Secretary of State's consent for the use of a floating installation in prescribed circumstances. The Regulations also require the Secretary of State's consent to the use of a mobile installation for the purposes of the extraction of petroleum where the main purpose of such extraction is the testing of a well.

For projects which fall outside Annex I of the Directive, the Secretary of State decides whether the project is likely to have significant effects and therefore whether EIA is required.

Nuclear Power Stations

Annex 1.2 of the Directive covers nuclear power stations and other nuclear reactors, and now includes their dismantling or decommissioning, so EIA is mandatory for these projects. It does not cover research installations for the production and conversion of fissionable and fertile material, whose maximum power does not exceed 1 kilowatt continuous thermal load.

For new nuclear power stations over 50 megawatts, the Secretary of State for Trade and Industry is the competent authority, and the Directive has been implemented in England and Wales by the Electricity Works (Assessment of Environmental Effects) Regulations 2000 (SI No 1927), which also applies to thermal power stations and overhead power lines. Power stations of less than 50 megawatts, and other buildings to house small nuclear reactors, are given consent through the normal planning system and are subject to the provisions described in Parts 1 and 2 of this booklet.

The Directive's provisions on the dismantling and decommissioning of nuclear power stations and reactors have been implemented by the Nuclear Reactors (Environmental Impact Assessment for Decommissioning) Regulations 1999 (SI No 2892) which apply to England, Wales and Scotland. They prohibit dismantling or decommissioning from being carried out without the consent of the Health and Safety Executive. A licensee who applies for consent is required to provide the Executive with an environmental statement. The Regulations also apply to changes to existing dismantling or decommissioning projects which may have significant effects on the environment (Annex II.13 of the Directive).

Other Power Stations and Overhead Power Lines

Annex I of the Directive includes thermal power stations with a heat output of 300 megawatts or more, and the construction of overhead power lines with a voltage of 220 kilovolts or more and a length of more than 15 kilometres.

The construction or extension of power stations exceeding 50 megawatts, and the installation of overhead power lines, require consent from the Secretary of State for Trade and Industry under sections 36 and 37 of the Electricity Act 1989, and the requirement for EIA has been implemented as part of the procedure for applications under those provisions. The Electricity Works (Environmental Impact Assessment) Regulations 2000 (SI No 1927) require EIA for all projects which fall within Annex I of the Directive. They also require proposed power stations not covered by Annex I, and all overhead power lines of 132 kilovolts or more, to be screened for EIA. Overhead power lines below 132 kilovolts are not normally expected to have significant environmental effects and do not need to be considered for EIA unless local circumstances demand it. The Regulations also make provisions relating to nuclear power stations.

The Regulations apply to England and Wales and replace that part of the Electricity and Pipe-line Works (Assessment of Environmental Effects) Regulations 1990 dealing with power stations and overhead cables. The 1990 Regulations, however, remain in force in respect of applications lodged before the new Regulations came into force. Power stations of 50 megawatts or less are approved under planning legislation and are subject to the provisions described in Parts 1 and 2 of this booklet.

FORESTRY PROJECTS

Projects in Annex II of the Directive include 'Initial afforestation and deforestation for the purposes of conversion to another type of land use.' The Environmental Impact Assessment (Forestry) (England and Wales) Regulations 1999 (SI No 2228) require anyone who proposes to carry out a forestry project that is likely to have significant effects on the environment to apply for a consent from the Forestry Commission before starting work. Those who apply for consent will be required to prepare an environmental statement.

There are four categories of forestry project that fall within the Regulations: afforestation (creating new woodlands), deforestation (conversion of woodland to another use), constructing forest roads, and quarrying material to construct forest roads.

The Regulations incorporate thresholds below which the project is taken to be unlikely to have significant effects on the environment. The thresholds are 5 hectares for afforestation and 1 hectare for other categories, although projects in sensitive areas have lower and in some cases no thresholds.

In considering whether EIA is necessary for projects which meet the appropriate threshold, the Forestry Commission takes account of the nature and scale of the project and the sensitivity of the site. Given the variability of forestry developments and the importance of location in determining whether significant effects on the environment are likely, it is not possible to formulate criteria which will provide a universal test of whether or not EIA is required.

The question must be considered on a case-by-case basis. Anyone contemplating a forestry project is advised to contact the appropriate Forestry Commission Conservancy Office for advice on whether or not EIA will be required.

An applicant who disagrees with the Commission's opinion that consent is required for a project may ask the appropriate Minister (in England, the Secretary of State for the Environment, Food and Rural Affairs) for a direction.

If the Forestry Commission discovers that work is being carried out that would have required its consent or has breached the conditions of consent, it may serve an Enforcement Notice. This can require the person carrying out the work, or someone else with sufficient interest in the land, to do one or more of the following: stop the work, apply for consent, restore the land to its condition before the work was started, carry out work to secure compliance with the conditions of consent, remove or alleviate any injury to the environment which has been caused by the work.

LAND DRAINAGE IMPROVEMENTS

Land drainage projects fall within Annex II. 1(c) of the Directive. New land drainage works, including flood defence works and defences against the sea, require planning permission and the provisions described in Parts 1 and 2 of this booklet apply to such works. Land drainage improvement works undertaken by drainage bodies are permitted development under the Town and Country Planning (General Permitted Development) Order 1995 and are therefore exempt from planning permission. As such works might have significant effects on the environment, the principles of EIA need to be applied to them. This is done through the Environmental Impact Assessment (Land Drainage Improvement Works) Regulations 1999 (SI No 1783), which apply to England and Wales and revoke and replace earlier Regulations which came into force in 1988 and were amended in 1995.

The 1999 Regulations require a drainage body to consider whether proposed improvement works are likely to have significant effects on the environment. Where the drainage body considers that there are unlikely to be such effects, it must publicise its intention to carry out the works. If the drainage body receives representations that there are likely to be significant effects but it still thinks otherwise, it must apply for a determination to the appropriate authority (Secretary of State for the Environment, Food and Rural Affairs or, in Wales, the National Assembly for Wales).

Where the drainage body concludes that the works are likely to have significant environmental effects, it must publicise its intention to prepare an environmental statement and notify specified consultation bodies (English Nature, the Countryside Agency and any other authority or organisation which might have an interest).

The environmental effects of the improvement works must be assessed in the light, in particular, of the environmental statement and any representations received. If there are no objections, the drainage body may determine that it will proceed with the works. If there are objections, the proposal must be referred to the appropriate authority for a determination giving or refusing consent to the works. The determination must be publicised.

Ports and Harbours

Annex 1.8 of the Directive includes trading ports, piers for loading and unloading connected to land and outside ports (excluding ferry piers) which can take vessels of over 1350 tonnes. Annex II.10(e) includes construction of harbours and port installations, including fishing harbours, which do not fall within Annex I.

Proposed development at ports and harbours down to the low water mark is subject to Town and Country Planning EIA Regulations. For other works, the requirements of the Directive have been implemented in England, Wales and Scotland by the Harbour Works (Environmental Impact Assessment) Regulations 1999 (SI No 3445). These Regulations incorporate a revised Schedule 3 to the Harbour Works Act 1964, which gives the procedures for harbour revision orders and harbour empowerment orders, and replace the Harbour Works (Assessment of Environmental Effects) (No 2) Regulations 1989 governing harbour works not needing to be authorised by harbour order.

Whether or not proposed works are subject to a harbour order, the appropriate Minister (generally the Secretary of State for Transport) is responsible for deciding whether EIA is required and, if so, whether the works should proceed in the light of the assessment and the comments of environmental bodies.

In general, works under harbour empowerment orders are likely to fall within Annex I of the Directive and require EIA in every case, whereas works under harbour revision orders are likely to fall within Annex II and the need for EIA will depend on whether there are likely to be significant environmental effects. Proposed works which are subject to harbour orders are not subject to EIA if the area of works comprised in the project does not exceed 1 hectare, unless any part falls within a `sensitive area' or the Secretary of State decides that EIA is needed in a particular case.

Marine Fish Farming

Intensive fish farming is an Annex II project. Developments of onshore fish farming facilities may require planning permission and the provisions described in Parts 1 and 2 of this booklet apply to proposals for such facilities.

Off-shore facilities do not require planning permission but require a lease from the Crown Estate Commissioners (or, where appropriate, from the Shetlands Islands Council or Orkney Islands Council). The Environmental Impact Assessment (Fish Farming in Marine Waters) Regulations 1999 (SI No 367), which apply to England, Wales and Scotland, set indicative criteria which, when triggered, require the Commissioners to screen applications to determine whether EIA is needed. They apply where the development is designed to hold a biomass of 100 tonnes or more, or where it will extend to 0.1 hectare or more of the surface area of the marine waters (including any proposed structures or excavations), or where any part of the development is to be carried out in a sensitive area. The Regulations prohibit a lease from being granted before the environmental statement provided by the developer, and comments on it from consultees and the general public, have been considered.

The Crown Estate Commissioners have, in consultation with the Scottish Executive Rural Affairs Department, issued an Environmental Assessment Guidance Manual for Marine Salmon Farmers which provides guidance to developers on how to conduct EIA and prepare an environmental statement. Guidance on the practical implications of the 1999 Regulations has also been issued by the Scottish Executive Rural Affairs Department.

Marine Dredging for Minerals

Extraction of minerals by marine dredging falls within Annex II.2 (c) of the Directive. The Environmental Impact Assessment and Habitats (Extraction of Minerals by Marine

Dredging) Regulations, which are likely to come into force early in 2001, will implement the Directive in waters around the UK (but excluding waters around Scotland). They will also implement the Habitats Directive (92/43/EEC).

Unless the regulator (Secretary of State in respect of English Waters or National Assembly in respect of Welsh waters) has provided, in response to a request, a written determination that proposed dredging is not likely to have significant effects, application for permission must be obtained and the application must be accompanied by an environmental statement. The only exceptions to this are where the Secretary of State has determined that the dredging would form part of a project serving national defence purposes, or where the regulator has determined that it is exempted under Article 2.3 of the Directive.

Transport and Works Act 1992

The Transport and Works Act 1992 (TWA) enables orders to be made authorising the construction or operation of railways, tramways, other guided transport systems and inland waterways; and works interfering with rights of navigation. Orders may also authorise ancillary matters such as compulsory purchase and creating or extinguishing rights over land. Applicants for TWA orders may apply at the same time for a direction that planning permission is deemed to be granted.

The procedures for making applications for orders under Part 1 of the Act are contained in the Transport and Works (Applications and Objections Procedure) (England and Wales) Rules 2000 (SI No 2190), which also implement the Directive. These Rules apply in England and Wales and replace the earlier 1992 Rules (SI No 2902) except for applications made and not determined before the new Rules come into force on 16 October 2000. In England, most applications for orders are determined by the First Secretary of State. Applications for orders relating wholly to Wales fall to be determined by the National Assembly for Wales.

The procedures include provision for opposed orders to be considered by way of a public inquiry, hearing or exchanges of written representations. Schemes that are identified by the Secretary of State or the Assembly as being of national significance are also subject to Parliamentary approval prior to a public inquiry being held.

PROJECTS ARISING IN SCOTLAND AND NORTHERN IRELAND

The arrangements for environmental impact assessment in Scotland and Northern Ireland are broadly similar to those applying in England and Wales, but are mainly subject to separate legislative provisions. The Environmental Impact Assessment (Scotland) Regulations 1999 (SSI 1999/1) implement the Directive in Scotland for projects which are subject to planning permission and for certain trunk road projects and land drainage works. Scottish Executive Development Department Circular 15/1999 explains the Regulations, and Planning Advice Note (PAN 58) on Environmental Impact Assessment also provides background information and guidance on best practice. The preceding section, of this booklet indicates which of those Regulations relating to projects outside the planning system in England and Wales, also apply in Scotland.

The Planning (Environmental Impact Assessment) Regulations (Northern Ireland) 1999 (SR 1999/73) implement the Directive in Northern Ireland for projects subject to planning permission. Development Control Advice Note 10 (Revised 1999) issued by the Planning Service, an agency of the Department of the Environment for Northern Ireland, explains the Regulations.

Appendix 8 includes a list of all implementing legislation currently in force in England and Wales, Scotland and Northern Ireland, and a list of available guidance. Go to table of contents

CONSOLIDATION OF DIRECTIVE 85/33 7/EEC ON THE ASSESSMENT OF THE EFFECTS OF CERTAIN PUBLIC AND PRIVATE PROJECTS ON THE ENVIRONMENT

As amended by Council Directive 97/11/EC (adopted by Council 3 March 1997). (Text of Directive 97/I1/EC is available in the Official Journal of the European Communities, 14 March 1997). Text in italics is a change or an addition made by 97/11/EC.

The Council of the European Union Having regard to the Treaty establishing the European Community, and in particular Article 130s (1) thereof, Having regard to the proposal from the Commission 1, Having regard to the opinion of the Economic and Social Committee 2, Having regard to the opinion of the Committee of the Regions 3, Acting in accordance with the procedure laid down in Article 189c of the Treaty 4.

1. Whereas Council Directive 85/337/EEC of 27 June 1985 on the assessment of the effects of certain public and private projects on the environment 5 aims at providing the competent authorities with relevant information to enable them to take a decision on a specific project in full knowledge of the project's likely significant impact on the environment; where as the assessment procedure is a fundamental instrument of environmental policy as defined in Article 130r of the Treaty and of the Fifth Community Programme of policy and action in relation to the environment and sustainable development;
2. Whereas, pursuant to Article 130r (2) of the Treaty, Community policy on the environment is based on the precautionary principle and on the principle(sic) that preventive action should be taken, that environmental damage should as a priority be rectified at source and that the polluter should pay;
3. Whereas the main principles of the assessment of environmental effects should be harmonised and whereas the Member States may lay down stricter rules to protect the environment;
4. Whereas experience acquired in environmental impact assessment, as recorded in the report on the implementation of Directive 85/337/EEC, adopted by the Commission on 2 April 1993, shows that it is necessary to introduce provisions designed to clarify, supplement and improve the rules on the assessment

procedure, in order to ensure that the Directive is applied in an increasingly harmonised and efficient manner;

5. Whereas projects for which an assessment is required should be subject to a requirement for development consent; whereas the assessment should be carried out before such a consent is granted;
6. Whereas it is appropriate to make additions to the list of projects which have significant effects on the environment and which must on that account as a rule be made subject to systematic assessment;
7. Whereas projects of other types may not have significant effects on the environment in every case; whereas these projects should be assessed where Member States consider they are likely to have significant effects on the environment;
8. Whereas Member States may set thresholds or criteria for the purpose of determining which such projects should be subject to assessment on the basis of the significance of their environmental effects; whereas Member States should not be required to examine projects below those thresholds or outside those criteria on a case-by-case basis;
9. Whereas when setting such thresholds or criteria or examining projects on a case-by-case basis for the purpose of determining which projects should be subject to assessment on the basis of their significant environmental effects Member States should take account of the relevant selection criteria set out in this Directive; whereas, in accordance with the subsidiarity principle, the Member States are in the best position to apply these criteria in specific instances;
10. Whereas the existence of a location criterion referring to special protection areas designated by Member States pursuant to Council Directive79/409/ EEC of 2 May 1979 on the conservation of wild birds 6 and 92/43/EEC of 21 May 1992 on the conservation of natural habitats and of wild fauna and flora 7 does not imply necessarily

that projects in those areas are to be automatically subject to an assessment under this Directive;

11. Whereas it is appropriate to introduce a procedure in order to enable the developer to obtain an opinion from the competent authorities on the content and the extent of the information to be elaborated and supplied for the assessment; whereas Member States, in the framework of this procedure, may require the developer to provide, inter alia, alternatives for the projects for which it intends to submit an application;
12. Whereas it is desirable to strengthen the provisions concerning environmental impact assessment in a transboundary context to take account of developments a tinternational level;
13. Whereas the Community signed the Convention on Environmental Impact Assessment in a Transboundary Context on 25 February 1991. (Preamblc of 85/337/EEC) The Council of the European Communities

Having regard to the Treaty establishing the European Economic Community, and in particular Articles 100 and 235 thereof, Having regard to the proposal from thc Commission 8, Having regard to the opinion of the European Parliament 9, Having regard to the opinion of the Economic and Social Committee 10, Whereas the 1973 11 and 1977 12 action programmes of the European Communities on the environment, as well as the 1983 13 action programme, the main outlines of which have been approved by the Council of the European Communities and there representatives of the Governments of the Member States, stress that the best environmental policy consists in preventing the creation of pollution or nuisances at source, rather than subsequently trying to counteract their effects; whereas they affirm the need to take effects on the environment into account at the earliest possible stage in all the technical planning and decision-making processes; whereas to that end, they provide for the implementation of procedures to evaluate such effects; Whereas the disparities between the laws in force in the various Member States with regard to the assessment of the environmental effects of public and private projects may create unfavourable competitive conditions and thereby directly affect the functioning of the common market; whereas, therefore, it is necessary to approximate national laws in this field pursuant to Article 100 of the Treaty; Whereas, in addition, it is necessary to achieve one of the Community's objectives in the sphere of the protection of the environment and the quality of life.

Whereas, since the Treaty has not provided the powers required for this end, recourse should be had to Article 235 of thc Treaty; Whereas general principles for the assessment of environmental effects should be introduced with a view to supplementing and coordinating development consent procedures governing public and private projects likely to have a major effect on the environment.

Whereas development consent for public and private projects which are likely to have significant effects on the environment should be granted only after prior assessment of the likely significant environmental effects of these projects has been carried out; whereas this assessment must be conducted on the basis of the appropriate information supplied by the developer, which may be supplemented by the authorities and by the people who may be concerned by the project in question; Whereas the principles of the assessment of environmental effects should be harmonised, in particular with reference to the projects which should be subject to assessment, the main obligations of the developers and the content of the assessment.

Whereas projects belonging to certain types have significant effects on the environment and these projects must as a rule be subject to systematic assessment.

Whereas projects of other types may not have significant effects on the environment in every case and whereas these projects should be assessed where the Member States consider that their characteristics so require; Whereas, for projects which are subject to assessment, a certain minimal amount of information must be supplied, concerning the project and its effects.

Whereas the effects of a project on the environment must be assessed in order to take account of concerns to protect human health, to contribute by means of a better environment to the quality of life, to ensure maintenance of the diversity of species and to maintain the reproductive capacity of the ecosystem as a basic resource for life.

Whereas, however, this Directive should not be applied to projects the details of which are adopted by a specific act of national legislation, since the objectives of this Directive, including that of supplying information, are achieved through the legislative process.

Whereas, furthermore, it may be appropriate in exceptional cases to exempt a specific project from the assessment procedures laid down by this Directive, subject to appropriate information being supplied to the Commission.

7

ENVIRONMENTAL DECISION MAKING

INTRODUCTION

Public participation practice is a growing part of environmental decision-making. This study looks at its genesis in the latter half of the 20th century. It emerged from an increasing awareness of the fragility of the biosphere in which we all live, coupled with a growing desire to get involved in its protection. Emerging from the Earth Summit in 1992, Principle 10 and Agenda 21 both called for increased public participation in environmental decision-making (EDM) and led to the adoption in Europe of the Aarhus Convention, to which Ireland is a signatory. The legislation resulting from Aarhus is discussed along with its implications. The pertinent legislation is the backdrop, but many other factors affect the development and effectiveness of participatory processes. These factors along with the costs and benefits are outlined. Taking all the relevant factors into account, it is clear that effective participatory processes in EDM require good planning.

The setting of objectives, process planning, selecting the relevant public, and choosing the appropriate methods or tools to use at each stage of a process, are all part of that planning. These are all described together with methods for assessing their effectiveness. In order to see how the methods explored might be applied to a particular process, Environmental Impact Assessment (EIA) is looked at in some detail. International best practice in public participation in EIA is also examined. Methods for assessing the state of public participation on a national level are considered, and a preliminary look is taken at Ireland's status, in the light of all of these. Finally, a survey of Local Authorities, Industry, Environmental Awareness Officers and the Public looks at; attitudes to, experiences of, capacity to deliver, and capacity building measures for, public participation in EDM. The overall aim of the study was to create a picture of public participation in EDM, as practiced world wide, and to look at what is possible in an Irish context. In order to fulfill the promise of Principle 10 in Ireland, it is necessary to develop effective, creative and worthwhile public participation practice in EDM. This treatise attempts to provide a starting point for the necessary debate required to bring about that promise.

PUBLIC PARTICIPATION IN ENVIRONMENTAL DECISION MAKING

The title of this study, poses several questions: What is the 'public'? What is 'participation'? What are the 'decisions' to which it refers? Starting with the last and working backwards, EDM refers to any process of decision-making where consequent significant environmental impacts are a possibility. This includes law making, planning, strategic planning, resource management planning, licensing of industry e.g. IPPC, environmental assessment (EIA), spatial planning etc. EDM can be even more complex than decision-making on other public issues.

First, environmental impacts do not respect property, jurisdiction or boundaries.

Second, EDM can involve government agencies as both manager and regulator.

Thirdly, environmental issues can provide especially heated value conflicts that require value trade offs.

The level at which the public is involved varies with the relevant legislation, and the attitude of the other stakeholders. Often it just means informing the public of a previously made decision and asking for comments, which may or may not be heeded. Sometimes it means informed consultation. Here there is exchange of information prior to the relevant authority's reasoned decision making, and all inputs are included and seen to be included. An example of this being the River Basin Management process under Article 14 of the Water Framework Directive as described by Judith Cuff (Cuff, J. 2001). Another example would be the Bantry Bay Coastal Zone Charter. Public participation, at its apex, could also mean that the public itself, in consultation with the relevant bodies, makes the final decision itself. Examples of this can be found under the auspices of Local Agenda 21. For public participation to be effective at any level, it requires the public to be well informed and kept aware of the possibility of participation. This requires a pro-active approach from industry and the relevant public bodies.

What then is 'the public'? The public is often treated as a unitary body, whereas in reality it is a collection of numerous continually shifting interests and alliances (Ortolano), which may be in conflict with each other. The term is used as a "catch-all to describe those with an interest in a decision, other than a proponent, operator, or responsible authority". (Petts and Leach). The individuals making up a public may be involved as individuals or as members of organisations. They may become involved due to their proximity, economics, social or environmental issues, values, etc. There is further discussion on this in Section 5. By contrast, stakeholders, of which the public is one, are literally those with a stake in an issue and may include non-governmental organizations (NGO's), government or its agents, industry, individuals, communities etc.

Stakeholders do not always want to be involved in an EDM process, but they have the right to know if their interests are affected. They may want to become involved at a different stake of the EDM process.

BENEFITS, COSTS AND THE FORCES AFFECTING PUBLIC PARTICIPATION IN EDM

Public participation in EDM is playing an increasingly important role in the protection of the environment. The forces driving the increased use of participatory practice, the benefits accruing from same, as will as the costs and potential problems are described below.

Forces Driving Stakeholder Processes

Many different forces are at work, prior to, and throughout, a public participation programme. However, the following factors have also moved the participation debate forward. People generally resist change, especially when they do not understand or agree with the goals, the methods, the sponsor or the timing of the proposed change. So, keeping the public in the dark is often a recipe for disaster. Citizens are also increasingly reluctant to defer to 'expert' government agency opinions, and are unwilling to act as sounding boards for bodies that have already made decisions, particularly when they affect their local communities.

Both of these tendencies lead to a lack of public confidence and trust, unless the relevant authorities take note of them. There are growing social expectations for environmental equality resulting in part from the high media profile given to environmental issues. Findings, that minority and low income groups in the USA had been disproportionately affected by pollution and downgraded environments, resulted in the issuing of Executive Order 12898 by President Clinton.

A far-reaching programme ensued, based on meaningful involvement and fair treatment for all races, cultures and incomes in environmental decision-making. "The transmittance of 'experience', rather than knowledge, is the critical neglected dimension of decision-making, particularly as it pertains to our national decision-making"(Barkenbus, J. 1998, p.9). For example development planning in Dublin City requires largely the same expertise as rural development planning, but the experience, vision and values of the rural population may be very different to that of their city cousins. The further the EDM process is from central government, the more pertinent becomes the public participation in same, and in this context, face-to-face meetings are critical to producing real learning and trust.

Previously, with the exception of Development Plans and Waste Management Plans, public involvement in EDM in Ireland was largely restricted to the minimal fulfillment of the two "rules of national and constitutional justice, and in particular 'audi alteram partem' (hear both sides)" (Scannell, Y., Cannon, R., Clarke, M., and Doyle, O. 1999).

In practice this meant that firstly a decision was made and this was then presented to the public for comment, a process also known as 'report and comment'. So, both sides have been heard, but one has the power, and can choose to ignore the input of the other. Changes driven by the Aarhus Declaration and Principal 10 will affect many areas of environmental regulation and planning, opening them up to far greater public participation. Alongside this is the increasing transparency of regulatory processes.

The directive on Freedom of Access to Information on the Environment (Directive 90/313/EC) SI.125 of 1998 and The Freedom of Information Act, 1997 have opened up the EDM processes to closer scrutiny by the public. By compelling the relevant authorities to release documents on request, the inner workings of government have been opened up for closer examination, and relevant information freed up to enable more effective public participation in environmental governance. There is also a growing diffusion of information. The many portals of information available on the internet, together with the traditional media, libraries and the excellent work, in Ireland, of ENFO, all increase awareness of environmental issues.

Since the Aarhus Convention, which agreed to recognise and support the crucial role played by ENGO's in articulating opinions of an environmentally concerned public, European funding has been given to support their activities (Jeffrey, D.W. 2001). □4.62 million was distributed in Europe in 2003. In Ireland, the Department of the Environment and Local Government allocated □132,000 in 2001, and □133,000 in 2002 to build capacity in 23 ENGOs with a national remit.

This was done via a limited company known as Environmental (Ecological) Core Funding Ltd. (Morrisey, P. 2003). It was estimated in 1999 that there were some 500 International ENGOs (Environmental NGOs) world-wide that operated in three or more countries. Apart from the rapid growth of the ENGOs, there is clearly expanded interest in participation. Public unrest at the prospect of development, without formal opportunity for participation in decision-making, is increasingly responsible for expensive delays. In the UK, policing of antiroad protests has alone cost millions of pounds (Treweek, J.1999). The Shell Oil Company, in the UK, was eventually forced to adopt a major consultation programme on the fate of the Brent Spar, North Sea oil exploration rig, as a result of the international reaction to the campaign waged by Greenpeace. (Acland, A. 2002, 4-1).

Hampshire County Council, in England, with a population of some 2,000,000, had to scrap its waste management plan, following much public unrest, and start from scratch, with a well planned and successful public participation programme. A waste plan was then adopted which had broad support, and is being put into action. (Acland, A. 2002, 4-2). Many industry organizations are realizing that for their good image, public safety, and the longterm financial well-being of their members, responsible partnership with host communities is essential.

There is some evidence that corporations which build sound, responsive relationships with all their stakeholders tend to be longer lived and more successful as businesses in the long term than other corporations (Wheeler and Sillanpaa, 1997). Olson, M. and Toyne, P.(2000) also report a study which suggests a direct correlation between increased public involvement of the public in corporate affairs, and increased profits.

The Chemical Manufacturers Association in the U.S.A. (representing 90% of all chemical production) set up the Responsible Care Initiative, which sets codes of practice. There are six areas covered by these codes viz: Community Awareness and Emergency Response (CAER); The Pollution Code; The Process Safety Code; The Distribution Code; The Employee Health and Safety Code; and The Product Stewardship Code. In relation to all of these, the guiding principles exhort their members "to recognize and respond to community concerns about chemicals and our operations, and to work with others to solve problems created by past handling and disposal of hazardous substances." The adoption, by some industries, of environmental management systems, such as I.S.O. 14001, and E.M.A.S (Eco-Management and Audit Systems) shows their belief that a transparent and responsive industry is also more likely to be a viable one. E.M.A.S. requires public reporting of environmental management and auditing. Internationally, trends towards decentralization in many countries, increases in the numbers of countries with representative democratic forms of government, and conditions laid down by donor agencies on development loans, e.g. The World Bank, are some of the international forces bringing increased public involvement in EDM.

Whilst it is not the intention of the author to go into detail on risk assessment and disaster management., it is worth noting that public participation should, under Council Directive 96/82/EC (Seveso ?) and the Council Decision 1999/847/EC, play an important part in both of these important areas. Wates, N. (2000) gives guidelines for organizing community risk assessment and disaster management programmes, whilst the "UK Guidelines for Risk Assessment and Management" (Defra UK. 2002) give a good introductory insight into the social aspects of risk, and in particular the factors influencing risk perceptions. These factors are listed here and are also relevant to other areas of EDM.

- o Risks which are involuntarily imposed (e.g. pollution from an incinerator) tend to be seen as less acceptable than voluntary ones (e.g. bonfires)
- o Unfamiliar risks (e.g. GMO's) tend to cause concern especially when it appears that they are poorly understood by science.
- o Activities associated with a dreaded form of death (e.g. cancer) are viewed with alarm.
- o Man-made or technological risks (e.g. nuclear power) are less acceptable than natural ones (e.g. Radon). o A risk involving a single large-scale consequence (e.g. civil aviation accident) causes more concern than risks which result in numerous small scale consequences (e.g. car accidents).

- o Alarm may be caused by risks when the consequences of exposure are delayed and cause hidden or irreversible damage (e.g. ionizing radiations).
- o Inequitable distribution of risks and benefits as a result of a particular activity is likely to make the risk less acceptable. An example here might be the remaining operative Chernobyl nuclear power stations. Employees, knowing the risks involved but needing to earn a living, would have a different view of risk to others living in the vicinity, and different again to those living 100 kms away, and relying on the electricity to heat their homes in temperatures of-30°C.
- o Activities posing a threat to certain groups such as children or future generations are generally more worrying.
- o Risks which are the subject of controversy and contradictory information generally cause concern.

Benefits of Public Participation.

The protection and enhancement of the environment is the main aim of most EDM processes, and public participation can improve the quality of both the process and the end decision. The participation process is also a learning one for all the stakeholders involved, especially if there is a free flow of information between the parties.

For the public it can often be the first experience of taking an active part in the democratic process. EDM processes benefit from the direct and immediate knowledge held by citizens and business, concerning environmental conditions in their communities and industries. Encouraging the public and other stakeholders to share their knowledge, with the regulatory authorities, fosters better-informed decisions and decreases the likelihood of environmental harm, whilst increasing project viability. Of 25 overseas projects sponsored and evaluated by the World Bank, 13 failed mainly through lack of local input (E.L.I. Research Report, 1997) (Technical Document 1, 1996). It should be realized that "wisdom is not limited to scientific specialists and government officials," (120 p.6) and that "rational analysis, carried on in ignorance of political reality, may well end up so divorced form social reality, as to be of little use to anyone." (Barkenbus, J. 1998, p.2) (Acland, A. 2002).

Public participation gives broader perspectives on a particular process, and early involvement gives: added time to study issues and develop the process; enhanced credibility of the decision making process; early identification of the diverse perspectives on the issues of concern; and the generation of solution options (E.L.I. Research Report, 1997). Better-designed projects, which avoid costly delays in appraisal and implementation, can also result from early and planned consultations and public participation (Bisset, R., 2000) (Acland, A. 2002, 4-1 to 4-5). For the proponent, early participation can have the added benefit of diffusing opposition to a project. If a broad based consensus is built, it can also lead to a public sense of 'ownership' (Acland, A. 2002, 4-2). The public's enthusiasm is potentially a powerful motivating force for the EDM process. Public input can also supplement scarce government monitoring, inspection, and enforcement, resources (Bisset, R. 2000).

Involvement in EDM is a learning experience giving the public insight into the governance process. If the experience is good, it powers the way for future co-operation. The converse can also be true! Experience has also shown that environmental issues are powerful catalysts of civic 9 participation and serve as a good incentive for citizens' action and responsible democracy (Technical Document 1, 1996). Executive Order 12898 introduced by President Clinton is a far-reaching programme, based on meaningful involvement and fair treatment for all races, cultures and incomes in environmental decision-making. It has done much to prevent the dumping of dirty industry on the marginalized members of the USA citizenry. The order requires state agencies, and in particular the EPA, to pro-actively seek out the disenfranchised members of communities likely to be affected by developments, or already living in depleted environments, and encourage them to become involved in the relevant EDM processes.

Costs of Public Participation

There is limited information on the relative costs of using the different methods of participation practice, although some attempt is made to quantify these in the section 5.4. which describes each method. Actual costings for specific processes are given in a number of studies (Petts, J. and Leach, B. 2000) (Cuff, J. 2001. pps 85, 89 and 97). Clearly deliberative processes, which are used to engage relevant stakeholders in debate, discussion and deliberation, could only be done as part of a major project. Expenses range from □20-40,000 (for a citizen jury with 16 participants for 5 days), to □200-300,000 (for 3 community advisory groups of 16 people over 6 months). (Petts, J. and Leach, B. 2000, P. 41).

However, extensive use of participatory processes in the 'Wise Use of Flood Plains' project showed costs of about 10% of the above. Most of this funding was used to employ facilitators (Cuff, J. 2001). The costs to members of the public in time and money, as well as emotional stress in situations where access to EDM is limited, do not appear to have been estimated previously. An attempt is made to gather some facts about the first two in this study. The intensity of the stakeholder process, including commitments of time, energy and money, and the often, uncertain results, has led to a burnout phenomenon amongst many participants, and from all stakeholder groups. (Yosie, T.F. and Herbst, T.D. 1998) Excessive raised expectations about participation can lead to frustration and anger with a breakdown in trust. It is essential then to be clear with all participants as to the limits of influence of the process. Where an infrastructure project crosses planning jurisdiction boundaries or where more than one regulator is involved, the costs to the public, state and proponents are multiplied. The increased time scale and expense involved in public participation is of particular significance in projects with a small capital funding. An international example of this would be the AIJ projects (Activities Implemented Jointly under the Climate Change Convention). These projects already require appraisal by two governments and then there is the extra of public participations costs. After this they must compete with traditional energy supply systems, where often no regulatory costs are involved.

Problems Encountered in Public Participation Practice

The main areas of difficulty affecting public participation are attitudinal, lack of capacity to deliver programmes, lack of clarity about what outcomes are possible and the lack of a legislative framework. Some aspects of these issues are addressed below. The difference between success and failure is not always clear. If the public participation process delivers a clear decision, which is then overruled by the legislature, is this a success? (E.L.I. Research Report 1997 p. 13). A good example here is the Bantry Bay Integrated Coastal Zone Management (ICZM) Project, funded for three years under the EU Life Programme. After three years of intense community activity and the drawing up and enactment of a text-book public participation plan, the Bantry Bay Coastal Zone Charter was created.

At this point, no further funding was available, and the Project was closed down. Clearly the process was a success, but the result not. There is a general lack of experience of participatory processes in Ireland, and very few stakeholders have experienced planned participation processes. There may be suspicion, cynicism, or enthusiasm, but there is unlikely to be previous training. Suspicion about participation could be from politicians, who may feel their power is being diluted, or from ENGOs that have very specific and fixed viewpoints. Cynicism could be from previous public participation processes which were poorly managed, leaving participants reluctant to try again (E.L.I. Research Report, 1997). Citizens are frustrated when they are treated as adversaries, rather than welcome participants in the EDM process........ they feel they have been invited too late in the process. Public 'hearings' often do not include 'listening'. Citizens efforts and ideas are not included in proposals and no reason given.

The resources available to proponents and government are overwhelming. Litigation and direct action at least get a reaction" (C.E.Q. 1997. p.18). Similarly, public hearings are criticised for: being held at times inconvenient for the public, establishing an atmosphere that inhibits dialogue, and conducting proceedings that intimidate the public (Lein, K.L. 2003, p.56).

The lack of a clearly defined purpose can also means that the role of the public is sometimes unclear not only to the public but to those delivering the process, with the consequence that the participation process chosen is often unsuitable for the particular EDM process (Breggin, L and Hallman, H. 1999. p.44). The making available of the necessary resources for a worthwhile public participation process, whether by the proponent or the regulator, by the commitment of personnel and finances, is essential. The implications for project proponents, of not having a properly planned and funded participatory EDM process, may prove much more costly, than dealing with the backlash. In many circumstances, the choice increasingly is not whether to involve the public, but how to get the best value from the chosen process (Petts, J. and Leach, B. 2000).

To avoid hearing from only the activist or the powerful elite, and in order to get the widest sweep of opinion and information, authorities must reach out into the community (Canter, L.W. 1996. p.592). In this regard some activities are best avoided, for example public meetings.

In the absence of trained facilitators these can often come down to 'he who shouts loudest wins!' Lack of technical support for the public, and difficulties in getting access to information can diminish the ability of the public to play a meaningful part in EDM processes. Public access to independent expertise in a particular technical/scientific field can prove difficult, particularly where the proponent of a project is a monopoly employer in the industry in the state. Faced with five weeks to object to a planning proposal, for example, this is a major hurdle. Added to this, it is the experience of the author that stakeholders with no technical/scientific background find it problematic to access, comprehend or evaluate data and information, especially under these short time constraints.

The feelings of powerlessness engendered by these hurdles, add to public perceptions of a lack of influence at both lower and higher levels of national and European government. "Consultations of interested parties.... can only ever supplement, and never replace, procedures and decisions of legislative bodies which possess democratic legitimacy" (COM (2002) 704. final p.4). This sense of powerlessness may be part of the reason why, even in cases where considerable energy and resources are expended, to identify individuals and groups, only a small proportion of the public ever attend participation programmes (Petts, J. and Leach, B. 2000. p.18). Public involvement may actually result in an increased level of conflict (134.p18). Becoming involved in a consultative process where the decision has already been made, or where the possible outcomes are not made clear at the outset, can lead to a great deal of frustration.

Anger may follow when the input of the public is ignored, especially following prolonged constructive engagement with an EDM process. Conflict can also occur between professionals. One source of such disagreements could evolve from the different emphases of 'scientific peer review' and 'social peer review'. The scientific peer review process is well established, and is essential to assess the technical information provided to policy makers. The parallel social peer review, designed to obtain societal acceptance and legitimacy for the decisions rendered, by contrast, does not have a set of recognised professional standards. (E.L.I. Research Report. 1997. p.14). It should also be remembered that whilst science may strive for the 'truth', the opinions of scientists are coloured by their values and beliefs. Scientists, engineers and other 'experts' should show respect in the face of 'emotional', cultural and traditional forms of wisdom. The perceptions created by a technocratic philosophy encourage NIMBY activism and further polarise communities.

The resulting deadlock frustrates the ability of planners to achieve consensus on initial issues and contributes to anti-participation attitudes. Therefore, rather than define 'NIMBY'attitudes as an irrational response to problems ordinary citizens cannot grasp, the solution is to encourage more participation not less (Lein, K.L. 2003. p.56). The misconception that the public are overly emotional, ill-informed and solely motivated by selfish interests, in their obstruction of facilities for the common good, leads to an information-based strategy. The assumption being that, 'if only the public knew the truth!' This assumes the existence of an objective truth to know (Petts, J. and Leach, B. 2000. p.9). The conflict that arises from having different bodies making decisions on different parts of the same EDM process can often confuse the public understanding and increase frustrations. For example the executive split of planning and environmental matters in some aspects of development between the Local Authority and the EPA. Public involvement in these circumstances must deal with two different agencies (Canter, L.W. 1996. p.592) (Gray, A.J. et al 2002. p.6).

Organisations resist change, and communication blocks within organisations can prevent them from making timely responses to their publics. The established routines and organizational systems of many bodies seek to promote the 'status quo'. Organisational structures which support sustainability, and therefore public participation, are a threat to 'command and control' style management systems. Managers, who have risen to power in this style of structure, frequently resist the transition to alternative structures, which embrace public consultation (Griffiths, A. 2000). Many opportunities for public participation are laid down in the environmental legal framework, as shown in Appendix 1. Where legislation does not specify early public involvement, then it is at the discretion of the relevant authority, and dependent on the opinion of the relevant public servant.

For example, in the development planning process, the Manager's reporting of the consultative process clearly includes the names and nature of all submissions together with his response to them (Section 11 of the Planning and Development Act 2000), whereas in making an Air Quality Management Plan (under Section 48 of the Air Pollution Act, 1987 and Section 102 of the Environmental Protection Agency Act 1982) there is no legal compulsion to show how any submission might have affected the final decision.The recent use by Cork City Council of Article 81 of The Planning and Development Act,

Regulations, 2001, to enter into a public/private partnership with a company previously refused planning permission to develop the same site (TP 02/26234), leaves the public only minimal scope for comment and no route to appeal any decision made.Under section 19 of the Protection Of The Environment Act, 2003, the making, variation or replacement of a Waste Management Plan, has been made an Executive Function. This has removed the immediate political accountability of the decision-making, and so removed another route for public input.

THE HISTORICAL PERSPECTIVE

General provisions for public access to information, participation and justice go back 14 centuries in Islamic Law, as recognized in many countries of North Africa, The Middle East and Asia. (Ahmad, A., 2002) but the International Boundary Waters Treaty, 1909, was the first global treaty to include strong provisions for public access to information and participation (Bruch, c. and Filbey, M., 2002). Broad international recognition of public involvement including provisions for transparency, participation and justice dates back 55 years. The Universal Declaration of Human Rights in 1948 provided the kernels for generalized access to information (Article 19) and Justice (Articles 8 and 10) as well as the right to associate (Article 20). In the West widespread public awareness of changes in the balance between human life and the environment emerged in the1960's, with the publication of Silent Spring (Carson, R. 1962), The Population Bomb (Ehrlick, P.R.,1968) and Small is Beautiful (Schumacher, E.F., 1973).

The immediacy of television portrayals of such disasters as the Torrey Canyon and "acid rain" (Jeffrey, D.W., 2001) made an even greater impact. In the USA as in much of the West, the 1960's and 1970's were marked by a participation movement (Breggin, L. and Hallman, H., 1999) which led to legislative changes. The Freedom of Information Act, 1966 gave access to data and information held by government agencies. The National Environmental Policy Act, (NEPA) 1969 amongst its measures required public review of Environmental Impact Statements (EIS). Previously the Administrative Procedures Act of 1946, directed government agencies to introduce a notice and comment stage into rule making procedures and to keep registers.

This Act also allowed for the right of appeal to federal courts (Ortolano, L., 1997) So together these three pieces of legislation provided the foundation for the involvement of the public in environmental protection in the USA, and NEPA has largely been the basis on which environmental impact assessment world-wide has been based. The voice of environmental concern was internationalized at the Stockholm Conference, 1972 (United National Conference on the Human Environment), drawing together both developed and developing nations, but in the absence of the Soviet Unions and most of its allies. For the first time environmental non-government organizations (ENGO's) played a important part in the deliberations.

Stockholm was the event that turned the environment into a major issue at the international level, declaring the right of people to live "in an environment of a quality that permits a life of dignity and well being". (GEO-3. UNEP., 2002) Amongst its impacts were:

- Some 50 governments together with the Organisation for African Unity (OAU) have recognized the environment as a fundamental right.
- During 1971-1975, 31 major national environment laws were passed in OECD countries, compared with 32 in the previous 15 years.
- By 1982 the number of countries with environment ministries rose from 10 in 1972 to 100.

Although the link between environment and development was made in Stockholm, it was the 1983 World Commission on Environment and Development (WCED) also known as the Bruntland Commission that first developed the idea of Sustainable Development, in its broadest sense. The WCED report in 1987 resulted from consultations at all levels of the global society, and attempted to balance the needs of future generations with those of our own. In essence it tried to promote both the inter and intra generational equity (GeO-3 UNEP., 2002). However, in its concluding statement, the commission stated that "existing decision making structures could not cope with the demands of sustainable development" (WCED.1987). The UN Conference on Environment and Development (UNCED or "The Earth Summit") held in Rio De Janeiro, Brazil, saw 176 government, 100 heads of state, 10,000 delegates, 1400 NGO's and 9,000 journalists gather in what is still the largest event of its kind ever (GEO-3.UNEP., 2002). The summit was preceded by global consultations with hundreds of thousands of groups and individuals.

Amongst its many achievements, the two most relevant here are, Agenda 21, and Principle 10 (P.10), of the 'Rio Declaration on Environment and Development'. Agenda 21, in its 40 chapters, lays a solid foundation for the promotion of sustainable development. Chapter 8, Section 1, calls on governments to seek broader public participation in policy making and decision making for sustainable development. Section 3 is devoted to the strengthening of the roles played by the principal social groups: women, children and young people, indigenous populations; NGOs; local government; workers associations and unions; business and industry; and the scientific and technological community (NCEDR, 2003) (Agenda 21, Chapter 8.).

Chapter 40 establishes two programme areas to ensure that decision making is based on sound information: namely, bridging the gap in availability, quality, coherence, standardization and accessibility that exists between developed and developing countries; and improving availability of information. Principle 10 states:

"Environmental issues are best handled with the participation of all concerned citizens, at the relevant level. At the national level, each individual shall have appropriate access to information concerning the environment that is held by public authorities, including information on hazardous materials and activities in their communities, and the opportunity to participate in decision-making processes. States shall facilitate and encourage public awareness and participation by making information widely available. Effective access to judicial and administrative proceedings, including redress and remedy, shall be provided".

It would be difficult to overstate the significance of Principal 10 in providing a mandate to regional and international governance development in the decade following the Rio Declaration, as well as to an underlying commitment motivating national law making. Since 1992 a raft of regional and international treaties have incorporated P.10, whilst addressing such issues as Desertification, Bio-Diversity and Clean Development Mechanisms. International organizations such as the World Bank, and to a much lesser degree the World Trade Organisation, have made steps towards making public consultation and involvement an integral constituent of their development loans.

In the context of this study and from an Irish and European perspective, the most important initiative evolving from P.10 is the 1998 UN/ECE Convention on Access to Information, Public Participation in Decision-Making and Access to Justice in Environmental matters. The ECE (Economic Commission for Europe) is a regional body of the United Nations and includes 55 countries. The European Community plus 39 other nations have signed the convention (also known as the Aarhus Convention), which it was ratified in 2001. The Aarhus Convention is based on the three pillars, ensconced in P.10 and elaborated in Agenda 21. The pillars are; access to information; public participation in decision-making; and access to justice.

The Aarhus Convention

The Aarhus Convention sets out to put a shape on the aspirations of Principle 10 (P.10). It encapsulates the three main pillars of P.10, access to information, public participation in decision-making, and access to justice, and sets the basic tenets for all European legislation on EDM. Pillar I-Access to Information. Article 4 of Aarhus sets out the general right of persons to gain access to existing information, on request. Pillar II-Public Participation in Decision-Making. Article 6 covers public participation (PP) in decisions on specific activities, e.g. on the proposed locating, construction and operation of large facilities. Article 7 deals with PP in the development of plans programmes and policies relating to the environment, which includes sectional or land use plans, environmental action plans, and environmental policies at all levels. Article 8, covers public involvement in the preparation by public authorities of laws and regulations.

In each of the Articles, 6, 7 and 8, early involvement of the public is encouraged. Articles 7 and 8 are less precise than Article 6. Article 6 provides for a high level of involvement adequately guaranteed by law (Stec, S. and Casey-Lefkowitz, S., 2000. p.86). Article 6 can apply to spatial planning decisions, development consents, operating permits, discharge permits, or any particular proposed activity where the decision making may have a potentially significant impact on the environment. Article 6 is not limited then, to activities where an EIA is required or to those activities listed in Annex 1 of the Convention.

Pillar III-Access to Justice

Article 9 allows the public to challenge "acts and omissions" by either public authorities or private persons, as applied to articles 4 and 6, as well as allowing them a review of violations of domestic environmental laws. In relation to Article 6, members of the public must have a significant interest {Article 9 (Section 2(a)} or maintain impairment of a right (Article 9 section 2 (b)). A level of standing is given to both individuals and NGO's. The provisions of an expeditious, free or very cheap, alternative review mechanism established by the courts is to be made available.

Publicity about these procedures and removal of barriers to access are also stressed. It is clear that in order for Pillar II to operate, Pillars I and III must also be well established, i.e. access to decision making cannot be effective without good information resources and the possibility of enforcement through access to justice. It should be noted that Article 3 reminds the parties that the provisions of the convention are only minimum requirements.

The Implementation of Aarhus

The implications for environmental governance under European environmental law and consequently that of the Republic of Ireland are far reaching as will be seen below. Pillar I-Access to Information. National Level. The Freedom of Information Act, 1997, (FOIA) gives access to all information held by the state, including information on the environment, other than the exceptions listed in the Act. ENGOs, however have told the author that they see S.I.125 of 1998 as a more effective tool. Directive 90/313/EC (on public access to information on the environment), implemented in Ireland as S.I. 125, of 1998, (The Access To Information On The Environment, Regulations, 1998.) has already been repealed by the European Commission and replaced by Directive 2003/4/EC, which must be brought into Irish law by 14th February 2005. The resultant changes include:

- Persons or bodies which perform functions by arrangement with a public authority and which effect the environment are themselves deemed to be public authorities for the purposes of the Directive (Article2, Section 2(c)), and so should make available environmental information held by or for them on request.
- Applicants for information do not have to state an interest.
- Information should be made available in the format requested (Article 3, Section 4).
- "Course of Justice" exceptions refer to information that may adversely affect the course of justice (Article 4, section2 (c)) rather than the previous "sub-judice" exceptions.
- Definite response times and appeals procedures are laid down. EC Institute Level.

Regulation (EC) 1049/2001 grants European citizens and residents a right of access to documents of the European Parliament, Council and Commissions. Applicants receive an acknowledgement of receipt, and within 15 days provide the item(s) requested or give reasons why not. The applicant is entitled to complain to the Ombudsman or institute Court proceedings if a negative reply is received. 20 pages of A4 text will be photocopied free of charge and any extra charged at cost.

Pillar II. Access to Participation

National Level

The proposal for a Directive of the European Parliament and Council providing for Public Participation in respect of drawing up of Certain Plans and Programmes relating to the Environment, and Amending Directives 85/337/EC as implemented in Ireland by S.I. 349of 1989 (and amended by Directive 97/11/EC), and 96/61/EC. These two refer to "Environmental Impact Assessment" and "Integrated Pollution Prevention and Control", respectively. The new directive will make changes in PP procedures required by these directives as well as creating a requirement to consult the public in drawing up of plans or programmes in seven other directives viz:

- The 1975 Framework Waste Directive (75/442)
- The 1991 Batteries Directive (91.157)
- The 1991 Agricultural Nitrate Directive (91/676)
- The 1991 Hazardous Waste Directive
- The 1994 Packaging Directive (94/62)
- The 1996 Ambient Air Quality Directive (96/62)
- The 1999 Waste Landfill Directive (99/31)

Honouring the subsidiarity principle, the proposed directive sets out the minimum requirement for effective public participation, in fulfillment of the objectives laid down by the EC Treaty in the field of the environment, leaving the detailed arrangements to the member states. The proposed directive requires member states to take the necessary measures, to ensure that the public are given early and effective opportunities to participate in the preparation (and review) of the plans and programmes listed above. For this purpose, "the public" has the same meaning as in Article 2(4) of the Aarhus Convention. To that end, Member States are required to ensure that:

- o the public are informed (by public notices or other appropriate means) about any proposals for such for such plans or programmes and that relevant information about such proposals is made available to the public.
- o the public are entitled to express comments and opinions before decisions on the plans and programmes are made;
- o and that in making those decisions, due account is required to be taken of the results of the public participation.

Taking the EIA directive 85/337/EC as an example, the following changes are proposed. The Aarhus Convention definitions of "the public", and of the "public concerned" are inserted into 85/337/EC. The public concerned will be defined as the public affected by or likely to be affected by, or having an interest in, the development consent procedure. A new Article 6(2) requires Member States to take steps to ensure early and effective opportunities to participate. The new Article 6(3) will lay down requirements for informing the public based on Article 6(2) of the convention. This requires the disclosure of all pertinent documents and opinions expressed by relevant public authorities. The new Article 6(4) requires Member States to provide proper opportunities to express comment and opinions as early as possible, but definitely before a decision is taken.

Reasonable time frames for the stages of participation should also be provided. Changes to Article 7 of the EIA Directive regarding Transboundary EIA are discussed. The IPPC Directive will have amendments similar to those of the EIA Directive, with the addition of a new Annex V in which detailed provisions for public participation will be set out. Both the Water Framework Directive (2000/60/EC) and the proposed Directive 2001/42 on "the assessment of the affects of certain plans and programmes on the environment (SEA Proposal), already include provisions which are consistent with Article 7 of the

Aarhus Convention (Banfield, N., 2003.). EC Institute Level Although no legislative proposal exists at present, COM 2002/704 Final, entitled, "Towards a reinfcrced culture of consultation and dialogue-General principles and minimum standards for consultation of interested parties by the Commission", provides good insight into the governance practice being developed within EC structures.

By fulfilling its duty to consult (Protocol No.7 annexed to the Amsterdam Treaty), "the Commission will help to improve the quality of policy outcome and at the same time, enhance the involvement of interested parties and the public at large". (COM.2002.p.5). The commission has underlined its intention to "reduce the risk of the policy makers just listening to one side of the argument or of particular groups getting privileged access" (COM.2002 p.5). These minimum standards laid down in the communication were themselves the result of a European wide consultative process.

Pillar III Access to Justice

National Level

Both Directive 2003/4/EC and the proposed Directive 2000/0331(COD) amending 85/337/EC and 96/61/EC, contain provisions for access to justice in line with the Aarhus Convention. Changes to the other directives listed are given in Annex 1 of the proposed directive 2000/0331(COD). However, shortcomings in the enforcement of environmental law throughout Europe, as outlined in the Sixth Environment Action Programme (6EAP), (Com (2002) 31, Final), will be the subject of a new Directive being worked on by the commission at present (Banfield, N.,2002).. All of these measures will become part of Irish Law in due course, when transposed by the Dail. In 6EAP it is recognized that better access to courts for NGO's and individuals would have a number of helpful effects in relation the implementation of community law. EC Institution Level.

Rules on access to justice with regard to community institutions are laid down in the Treaty of Rome. The Aarhus

Convention will have to be ratified by a legislative instrument, taking account of existing provisions, in order to ensure compliance at Community Institution level. In Johannesburg, South Africa 2002, a new initiative called Partnership for Principle 10 (PP.10) was launched by governments, donor groups and NGO's, as one of the outcomes of the World Summit on Sustainable Development (WSSD) (WRI Press release, 2002). In UN parlance this is a Type 2 outcome, meaning it involves non-governmental stakeholders, whilst being linked to inter governmental commitments, in this instance Principle 10, (P.10) and WSSD Plan of Action.

In PP.10 each partner makes specific commitments based on the promotion of P.10 at national and international level, by identifying implementation gaps, by capacity building, and by improving and implementing relevant legislation. Each partner is held accountable for fulfilling its commitments. PP.10 strategy is to support independent assessment by NGO's of their countries performances based on a common framework. Independent assessment by NGO's using systems such as The Access Initiative (Petkora, E., Maurer, C., Henninger, N. and Irwin, F.,2002.) are to be received by government agencies and other stakeholders and priorities for action identified. The European Commission has committed itself to PP10.

Transboundary Environmental Impacts

The Convention on Environmental Impact Assessment in a Transboundary Context (the ESP00 Convention), was signed by the European Community in 1991. This initiated changes incorporated into the EIA Directive 97/11/EC and the IPPC Directive 96/61/EC, creating provision for cross border public consultation. Article 7 of the EIA Directive requires member states to send information 'as soon as possible' and not later than when its own public is informed. The affected state must be sent a designation of the project together with any available information on its possible transboundary impact, as well as information concerning the nature of the decision to be taken. If the affected state wishes to participate in the assessment process, it must also be sent the application submitted by the developer under Article 5, together with the information relevant to the assessment procedure. All the above information must be made available to the public in the affected State, should the latter decide to participate in the assessment process.

Within a reasonable time scale, submissions must then be accepted by the competent authority in the Member State, in which the development is proposed. The notifying Member State under Article 8 must take the transboundary consultations into account in making a decision, and under Article 9 must inform the Member States consulted under Article 7 of the decision made. With regard to the decision there is no compulsion to notify the public in the affected member states, even if they make submissions to the transboundary EIA process. (Macrory, R, and Turner, S., 2001). Article 17 of the IPPC Directive, drafted in parallel with the 97/11 Directive is less explicit with regard to the provision concerning cross border public consultations and it is proposed to amend both to bring them in line with the Aarhus Convention.

With regard to transboundary EIA, these proposed amendments will ensure equal rights of consultation for the public whether they live in the notifying or affected Member States. However, in the case of the IPPC permitting process, the public of the affected state will remain entitled only to access the IPPC application itself (Macrory, R. and Turner, S. p.14). The Water Framework Directive is still the only measure that includes not just transboundary consultations but also power sharing (Article 14(i).) Regardless of the above, it would appear that Ireland had failed to implement fully the 1985 and 1997 EIA Directives with regard to Transboundary consultations, as ruled by the European Court (Case c-392/96 (1999) CMLR 727). The Planning and Development Act 2000 whilst it empowers the Minister for the Environment to make regulations regarding transboundary EIA, it 26 does not give rights of consultations for the Irish public nor that of other Member States (Macrory, R. and Turner, S. p.17). As can be seen from the above, there is a great lack of clarity on the issue of consultation with regard to transboundary environmental issues.

PUBLIC PARTICIPATION PRACTICE

Programmes to encourage the public in EDM processes often have multiple purposes, and no simple formula exists for success. In each case a programme has to be designed to fit the particular combination of process, proponent and public. A number of case studies are given by Acland, A. (2002. pps 25-30), Ortolano (2997, pps.409-417) and the European Commission (COM (2002) 704 final. pps 62-72) and are illustrative of this point. The proponent of an EDM process be it a public agency or private enterprise must be clear as to their aim in involving the public.There may also be different aims at different stages of a process, for example within the EIA process from the elicitation of values relevant to site selection at the project design stage, to the optimisation of trust and credibility at the monitoring stage (Petts, J. and Leach, B. 2000. p.20). In order to put together the aims with the methods it is necessary to have a plan:

Planning for Public Participation

In order to make a coherent plan it is necessary for the practitioner to establish the objectives of the process, based on the overall aims. It is then important to identify which public(s) the participatory process is trying to include.

Objective Setting

Clear objectives in planning a public participation process are essential, and ideally they are delineated by consultation with a representative group of stakeholders. A community assessment process, involving community interviews (Marxen, J. 2001. 6-4), would provide a good basis for a plan. The California Department of toxic Substance Control, in its 'remediation and permit procedures' involves a public participation specialist to review the findings of such interviews (Marxen, J. 2001. 6-20). It is equally essential that all those involved, both active and nonactive participants understand the objectives, as well as the limitations on what can be achieved. The connection between process and outcome must be clear to all. Ongoing process evaluation is essential to keep a clear focus on the objectives

(Cuff, J. 2001. p.75). The two basic reasons for identifying and classifying objectives are; that objectives change over the various stages of a process, and that some participation techniques are better than others for achieving particular objectives (Canter, L.W. 1996. p.593).

Important Factors in Objective Setting

Apart from the above, there are a number of specific factors that underlie the setting of objectives, that will need to be considered at the outset, and will be present throughout the EDM process. Environmental Protection The main purpose of EDM, and thus the main purpose of public participation in EDM is to achieve protection, conservation, and wise management of the environment. This can only be achieved if the proponent properly collects (and acts upon) evidence, opinions and perspectives from all the interested or affected stakeholders, who are to be fully involved in the EDM process, and from the earliest opportunity. Legal Requirements Appendix 1 shows the major current legislation relating to public participation in EDM in Ireland.

Opportunities to participate are present throughout the process of making a Development Plan, a Waste Management Plan and a River Basin Management Plan, however, most of the other processes are of the 'report and comment' type where the public's reaction is sought to an already drafted document. As discussed elsewhere, this situation is changing with the Directives that are emanating from Europe implementing the Aarhus convention. The legal processes, their limitations and the concomitant responsibilities should be made clear to all the participants at the outset. Proponent Objectives vs Public Objectives. The various stakeholders in an EDM process can have very different expectations of that process. The proponent may be only conducting a public relations exercise or fulfilling a legal requirement, having a final decision already made. The participants may want an influential voice in the EDM process and expect their input to be valued and included.

The proponent, seeking a mandate for a particular project, may seek to defuse opposition through consensus building.

Conversely, the individuals or groups involved may be largely uncompromising on their expected outcomes. The proponent may be restricted by time and/or money, whereas the public participants may be highly motivated by value judgments made on a particular issue, and prepared to commit much personal time and energy to it. Participation can then be viewed as a tool that the public and the project proponent or planmaking agency can use for their own purposes. Two Way Communication. The development of a two-way process of communication between the regulatory authority and the public is essential in order to identify public concerns and values; inform citizens of authority's plans; and inform the authority about alternatives and impacts.

Experts vs. Public. Accepting that the role of regulatory authorities is to serve the public interest, the challenge is to integrate technocratic and democratic contributions, when addressing complex issues (Lein, K.L. 2003. p.57). Dearing and Rogers (1996) state bluntly that, "scientific research results do not play a role in the agenda setting process". The social construction of reality is what gives issues their importance, such that the building of a waste incinerator creates much stronger reactions and media attention that regular back yard burning of waste, regardless of the relative levels of dioxin release. Levels of Participation. The setting of objectives is constrained by the level of participation being allowed in any one process. It is useful to look at "Arnstein's ladder" (Ortolano, L. 1997. p.417) /and for the agency or proponent to analyse its motivations in establishing the process.

1. Manipulation
2. Ther Non-Participation. These levels assume a passive apy audience, which is given information that may be partial or constructed.
3. Informing
4. Consultation Degrees of Tokenism People are told what is about to happen, what is happening now, or what has already happened.
5. Placation People have a voice, but no power.

6. Partnership People's views have a small influence on decisions made by traditional power holders.
7. Delegated Power Degrees of Citizen Power People negotiate with traditional power holders, agreeing rules, roles, responsibilities and levels of control.
8. Citizen Control Some power is delegated.

Arnstein represents the levels of citizen participation as the 8 rungs of a ladder and groups the rungs into three groups. The continuum stretches from going through the empty ritual of nonparticipation, to having the real power needed to affect the outcome of the process. Arnstein describes this first category non-participation as tactics whose real objective is "to enable power holders to educate, or cure the participants". In the degrees of tokenism Arnstein argues that when these are "proffered by power holders as the total extent of participation, citizens may indeed hear and be heard. But they lack the power to ensure that they are heeded".

The final category, citizen power, involves citizens-agency partnerships such as those in the Bantry Bay Coastal Zone Charter and finally programmes in which citizens are in control, or can veto agency decisions.

The further the process is from centralised power, the further up that ladder it is possible to go. Thus, at the European Union level, public participation has reached its peak at the level of consultation, whereas there are now many local partnerships under Agenda 21. Generally, however, regulatory authorities are not empowered to go beyond the consultative level in public participation. Wilcox, D. (1994) adapted Arnstein's Ladder to give four categories of participation methods, and these present an easier model for the purposes of this study, as will be seen later. They are:

1. Education and information provision
2. Information feedback
3. Involvement and consultation
4. Extended involvement

These categories are explained in Section 5.4. To enable the selecting of objectives, Canter, L.W. (1996) looks at a number of systems for categorising_them. One such gives three general types of objectives, which further divide to give seven more specific, viz: Public Relations-Legitimizing the role of the proponent or regulatory body. Developing confidence and trust in the stakeholders. Information-Diagnosis of problems and needs. Development of alternative solutions. Evaluation of consequences and alternatives. Conflict resolution-Consensus seeking. Depolarizing interests. One international body, The Institute for Participatory Management and Planning (Connor, D.M. 2003) (I.P.M.P. 2003) goes much further in breaking down the objectives, with fifteen basic objectives under three categories.

The IPMP list is based on enabling a public participation practitioner, working for a regulatory authority or proponent for a project, to analyse their objectives, and then establish the best methods to achieve them. Each objective is coupled to a check-list. Viz:

- Responsibility. Establish the legitimacy of your Agency and you project.
- Maintain the legitimacy of your Agency and your project.
- Establish the legitimacy of your process.
- Maintain the legitimacy of your process.
- Establish and maintain the legitimacy of assumptions and earlier decisions.
- Responsiveness. Get to know all your potentially affected interests.
- Get to see the project through their eyes.
- Identify all potentially relevant problems.
- Generate solutions.
- Identify and classify the key issues.
- Effectiveness. Nurture and protect your credibility.
- Have your communication received and understood.

- Receive and understand information that's communicated to you.
- Search for common ground between polarized, potentially affected interests.
- Mediate between polarized interests.
- It is clear from this that one of the main objectives of practitioners is to maintain their credibility and that of the process, without which all else is in vain.

The code of ethics for practitioners, from the International Association for Public Participation Practitioners (IAP2), is also a useful general guide in setting the objectives of an effective process (IAPZ. (a). 21/5/03).

- o Purpose. We support public participation as a process to make better decisions that incorporate the interests and concerns of all affected stake holders and meet the needs of the decision making body.
- o Role of the Practitioner. We will enhance the public's participation in the decision-making process and assist decision makers in being responsive to the public's concerns and suggestions.
- o Trust. We will undertake and encourage actions that built trust and credibility for the process among all participants.
- o Defining the Public's Role. We will carefully consider and accurately portray the public's role in the decision making process.
- o Openness. We will encourage the disclosure of information relevant to the public's understanding and evaluation of a decision.
- o Respect for Communities. We will avoid strategies that risk polarising community interests, or that appear to 'divide and conquer'.
- o Advocacy. We will advocate for the process, and not for any interest, party or outcome.

o Commitments. We will ensure that all commitments made to the public, including those by the decision maker, are made in good faith.

Identifying the Public

There are as many publics as there are different people who care, positively or negatively, about a project. Who they are may depend on their ethical, moral, interest, welfare etc. viewpoints. Other reasons the public may be affected by regulatory decisions and plans include:

o Proximity-Pollution, Property values, Employment.

o Economics-Landowners, House-owners.

o Use-Amenity value, rights of way, vista.

o Social and Environmental Issues-Justice and Risk.

o Values-Animal Rights, Ecology, Religion.

These members of the public could be described as 'stakeholders' alongside the proponent, regulatory bodies, industry and NGO's. Wates, N. (2000 p.15) defines stakeholders as "persons or organisations with an interest, because they will be effected and may have some influence."

Whatever kind of process is chosen, it is essential to define which members of the public are stakeholders (Yosie, T.F. and Herbst, T.D. 1998. p.19). Different stakeholders are affected to varying degrees, and the greater their interest in a project, or the greater the potential impact on them, then the more involvement they should have. There are also those who could affect or impact on the project.

An individual's interest may change over the course of the participation process, and so they may opt in or out at various stages, others may want to be involved throughout the process. Some may purposefully reject participation. It is essential, however, to ensure that no group of persons is excluded, as this, whether intentional or not, will be resented, and may lead to action being taken in the legal or political spheres. The number of potential public stakeholders on some issues is practically limitless, however, not everyone can sit around the same table.

The challenge then is to balance the need to consider the many view points of all the stakeholders, with the practical considerations of convening a group of individuals who have a role in making, or directly influencing, decisions.

'Tiered' stakeholder involvement is one approach used to strike that balance, and in it three categories are defined:

o Those who want to be directly involved in the process

o Those affected who just wish to be kept informed

o Anyone with an interest in the project

Many stakeholders are not able to effectively participate in deliberations even when they are the most potentially affected. This may be due to poor language skills, poor self-esteem, lower educational achievement, physical handicap or intellectual disability and so on. Many remain unaware of their opportunity to participate.

The environmental Justice Programme of the US EPA, attempts to address these issues requiring "the decision makers to seek out and facilitate the involvement of those potentially affected". (US. E.P.A., 17/11/02). Participation specialists distinguish three ways of identifying segments of the public (Ortolano, L. 1997. p.405).

o Self identification

o Staff identification

o Third party identification

In Self Identification, individuals and groups come forward as a result of publicity by the agency or body conducting the participation programme. These are "active" responses and should be included in a mailing list and kept continually informed of the process. Staff identification-regulatory agency personnel learn from experience the names of individuals and groups that might be interested in involvement. GIS (Global Information Systems) are being used increasingly by agencies to enable both the public and themselves to ascertain the locality of environmental problems, as well as that of the stakeholders geographically relevant to them (US, E.P.A. 2000).

Other staff identification techniques include: Electoral Registers; Mailing Lists; analysis of local newspaper stories; local organisation lists; and maps. Third party identification can be achieved by asking those already involved about other individuals or groups that should be sought out. Local officials in other agencies can be asked to suggest people who may want to be involved. As each person becomes involved, they are asked to suggest others, until the response becomes minimal.

Various public involvement groups check lists exist, and can be useful as another way to ensure that no group is overlooked. (Canter, L.W. 1996. p.598) (DOELG, 1995. p.16) US E.P.A. 1999. 2-1). But it is particularly important to bear in mind that in some areas of cities, large ethnic or immigrant groups do not belong to listed groups, and are not used to interacting with government agencies. They may even be fearful of so doing. A social profile, drawn up by an applied sociologist would give a clear insight into the community from which the stakeholders are being sought. A community assessment process such as that used by the California Department of Toxic Substances Control (Marxen, J. 2001. 6-4 to 6-15) would also be an effective tool here.

Public Participation Techniques

In the Republic of Ireland, legislation has generally required the proponent of a project, or applicant for a license to notify the public of her/his intentions in this regard. The regulatory authority then takes responsibility for conducting the consultative process. However, although at present the proponents are only 'encouraged' (for example in the EPA guidelines for EIA), to involve the public at the earliest stage of project design, the Aarhus convention (Sections 4-6) 'requires' it.

So, for the future, a whole range of public and private bodies will be involved in public participation, and in an increasing variety of circumstances. The following list of good practice principles (European Commission, 2000. p.239) should be considered before selection of specific methods.

General Guiding Principles

- o Plan early for public participation to establish the resources needed to support the process, and identify the limitations of finances.
- o Identify stakeholders and their legitimacy and/or representativenesss.
- o Identify the appropriate techniques to be used for each stage of the process, and provide information in a form that is comprehensible to the recipients.
- o Hold events at a time and venue to suit all participants.
- o Allow sufficient time for stakeholders to assimilate and respond to information.
- o Ensure that the inputs of stakeholders are integrated into any decisions made.
- o Ensure feedback on all issues raised.

Which methodology to use depends on many factors, but the following questions adapted from (Cuff, J. 2001. p.18) (Acland, A. 2002. 3-2) will help in deciding.

- o How complex is the issue to be discussed?
- o What processes are already in place for resolving the issue and how well are they working?
- o What processes have been tried previously and how well did they work?
- o What is the geographical scale of the issue?
- o At what levels of societal structure is the process to operate?
- o What resources are available in time, money and in-house skills?
- o Will external professional facilitators or consultants need to be employed?

With regard to this last question, apart from the costs, the employment of consultants can raise other issues. Consultants are only as good as their brief, and whilst the process may work well, (Cuff, J. 2001. p.131) the end product may be less of a success. The use of professional facilitators combined with stakeholder training in facilitation skills, early in the process, would enable a wider range of techniques to be used. It would also make the use of group-work, one of the most useful tools in concensus-seeking, more viable. (Cuff, J. 2001. p.49).

Good Practice Principles Specifically For Project Proponents (Acland, A. 2002, 3.2) include:

- As soon as possible focus on specifics, away from slogans.
- Acknowledge differences-then keep an open mind, thinking creatively to build on common ground.
- Uncertainty leads to distrust to fear to hostility, so clarify any uncertainty as early as possible. Be as open as possible and explain why, when you cannot be.
- Where legally possible, acknowledge past mistakes.
- Resolve internal issues prior to public participation.
- Address all dimensions of a situation; differences of needs and interests, behavior, and beliefs and values.
- Only make commitments you will be able to keep.
- Being open, consistent and honest in stakeholder relations is essential to building of trust.
- Flexibility in the choice of methods and use of same enables the creativity of the process.

Most important of all, it should be remembered that, stakeholder dialogue is often messy, disjointed and even chaotic at times. This is just how it should be-reflecting the real world in which we all have to operate (Acland, A. 2002, 3-2). So events should be made enjoyable. Having regular breaks for stretching, food and refreshments.

Process Planning

Many of the questions posed above can be answered during a process planning session (Wates, N. 2000. p.106) which allows stakeholders to determine the most suitable public participation methods for their particular situation.In order to do this, it is necessary for the participation coordinators to make an early attempt at identifying key stakeholders.

Record Keeping and Reporting

In order to maintain continuity and for confidence building, records of all sessions/meetings are kept and reported to all stakeholders, together with outcomes from the process. This is essential.

Public Participation Tools

What follows then is a survey of some of the tools most widely used in public participation. They are arranged in groups based on their general purpose and according to the four levels of participation outlined by Leach, S., and Wingfield, M. (1997, p.27).

Level 1 Education and information provision.

Level 2 Information and feedback.

Level 3 Involvement and consultation.

Level 4 Extended involvement.

These four levels categorise methods, according to the amount of power transfer from the responsible authority to the public. In level one, the public is passive, receiving information. In level two, feedback is requested and considered as part of the EDM process. Level three involves information, sharing and consultation, on all relevant aspects, before the final decision is made by the authority. At the fourth level, the authority can delegate power to a group of individuals to make decisions on its behalf. Clearly, due to legislative restrictions, level four processes will be rare, except in a very local context. By contrast levels two and three will be widely used in the future. These levels are not, however, prescriptive, and some of the tools, such as interviews, can be combined with others at different levels. For a more detailed breakdown of the effectiveness of the different methods, see Canter, L.W. (1996).

The public participation tools described here are an amalgam of the author's experiences, and information from the following sources: (Canter, L.W. (1996), Gray, A.J. et al (2002) Yosie, T.F. and Herbst, T.D. (1998), IAP2(6) (2003), Ortolano, L. (1997), Connor, D.M. (2003), I.P.M.P. (2003) and Wates, N. (2000).

Level 1. Education and Information Provision

Printed Materials-Fact sheets, Newsletters, Brochures, Issue papers. The guiding rule here is KISS-Keep It Short and Simple, and embellish with good visual impact. The contents should recognise the pros and cons of an issue and explain the public's potential role in, and affect on, the project. An easy response method such as a postage paid card or free-phone hotline number will enable identification of stakeholders interested in involvement.

These materials can be targeted or general, and may reach a large audience. They may not, however, be read, and are limited in what information they can convey. Connor, D.M. (2003) estimates only 10% of the necessary information can be conveyed in this format. It is also difficulty to know if you have reached your audience, unless responses are generated.

Advertisements

Whilst advertisements in newspapers or magazines can potentially reach wide audiences, they are generally expensive and limited in the information they provide. It is essential to choose the pages, days and publications very carefully. Different publics read different publications! It is, as with other printed material, difficult to know if it has been read. Some feedback mechanism should be included.

Press Releases

The potential for reaching the public is great if the issue is of sufficient interest to the media, and can include requests for feedback. There may be problems of misrepresentation and lack of editorial control. The use of publicity stunts may be of use in drawing media attention to the less exciting areas of EDM. This is a cheaper option than advertising, though local newspapers will often do features on regular advertisers more readily.

Public Displays

Exhibits in public places have the advantage of enabling a wider public to view models and displays at their own choice of time and at their leisure. They are cheap in public buildings, but 42 to be more effective require the back up of personnel to explain details and/or provide literature. They may also need to be duplicated or travel around the stakeholder constituency. Newspaper Inserts. To be successful an insert needs to be: of the best graphic design, in order to catch the reader's attention; in the publication on days when others are not; and preferably in the relevant local paper, where it is likely to have greater impact.

Newsletters

Although everyone will not read a newsletter it is an ongoing source of information with enough print space for some detail. It can also be used as an insert in the local paper. A newsletter also has great potential for feedback, via reply paid response forms, with for example prize draws attached.

Bill Stuffers

Any of the printed formats above could be included with utility bills, rates bills etc. Feedback possibilities can be included, these to be returned along with the bill payment.

Information Repositories

Usually, the regulatory authority makes EDM documents available at its offices and during office hours. This excludes, or makes difficult, access for many stakeholders. Libraries, as an alternative, are locations which have hours designed for public accessibility and whose 'raison d'etre' is in part the making available of information (EC1, 2000). Information repositories are, however, not widely made use of by the public. (IAP2 (6), 2003).

Site Visits

Small groups of stakeholders can learn more from first hand experiences than from any other means. It may, however, be difficult to identify a suitable site, that replicates all the issues faced by those in the EDM process.

Video

A production controlled by the proponent or the regulator may be seen as propaganda, but if produced with stakeholder involvement may be a good substitute for site visits, and can be shown widely over the stakeholder constituency. Whilst it is expensive to produce, a professional and credible video, once made, is not too expensive to reproduce, and so can be distributed to community groups, NGO's, employees etc.

Independent Technical Documents and Expertise.

Such documents could give credibility to a process, though they may be incomprehensible to the general public unless a non-technical summary is given. Public access to experts builds credibility and may help resolve disputes over disputed 'facts'. The reality exists, however that these technical experts may contradict information provided by the project proponent.

Level 2. Information Feedback

Public Meetings

This widely used method is described by (Connor, D.M. 2003) as, "the last of the blood-sports, and like the others, should be banned". Public meetings usually involve a speaker or panel giving a presentation, followed by a question and answer session. Sometimes these are broken down into smaller group discussions returning to plenary sessions. Gray, A.J. et al, (2002) suggest that public meetings offer an opportunity for stakeholders to meet, as well as demonstrating the willingness of the proponent to meet the other interested parties. However, they go on to say that they are one of the most complex and unpredictable methods. They can also be intimidating and can be hijacked by vocal individuals or interest groups. Badly handled, they can entrench opposition to a project, and division within a community. As a means of sharing information with those able and interested to attend, it is reasonably cost effective if run by well-trained staff.

Public Hearings

These formal occasions have a hearing officer, and a stenographer. Formal presentations from the public are received. This rather rigid format should according to (IAPZ (b) 2003) be avoided if possible, as it does not foster constructive dialogue and can perpetuate an 'us versus them' feeling, with little dialogue between participants. This format enables the public to speak without rebuttal, which is positive or negative depending on perspective.

The Internet

The use of the internet as a means of making available large amount of regularly up-dated reference material has been seen by many as the way forward in information provisions. The development of the European Commissions 'single access point', 'Your Voice in Europe' (Europa, 2003) is a good example. This portal site provides access to: every information source in the EU; consultations, where opinions are solicited on a wide range of policy-making and rule-making decisions; discussions with the EU leaders on-line; scrutiny of the working of the EU, with the possibility of giving anonymous feed back; and useful links to other sites and ways the public can be heard. There are, however, pitfalls in relying too heavily on this low cost innovative tool, and its potentially global audience. (Hacklay, M. 2002) found that only 10% of the total population of the U.K. had daily access to the Internet, with only 3% of those coming from the lowest income group, and 48% from the uppermost group. Hacklay also found, that only a small minority of the population, having access to the Internet, regularly entered the environmental information sites. This minority came mainly from the educated middle classes. He also concluded that the traditional media were the most used sources of information on the environment.

The Central Statistics Office, 'Fourth Quarter 2000 Survey' (C.S.O., 2001) showed 417,000 households in Ireland owned computers, with 262,700 having internet connections. This latter figure representing 20.4% of all households was the quadruple of the 1998 figure, and is likely to be much lower than that existing in 2003. Based on these figures, and Hacklay's findings, it is clear that caution should be used in relying on the internet as a means of communication.

Free-phone Telephone Lines

Well advertised, these phone lines can be used for the public to obtain information, ask questions or make comments about proposals or issues. The person on the line must be dedicated, well trained and have access to the information themselves, with a guaranteed call back time for complex answers that require research. This shows transparency and allows for up to date information to be easily available. Good examples of this type of service are: the Europe Direct, Free Phone 00800 678910; and the ENFO information line, Lo-call, 1890 200191.

Interviews

One to one meetings with stakeholders provide for a two-way flow of in-depth information and can be used to evaluate potential members of community advisory/liaison groups. Interviews are very useful for developing and refining public involvement and consensus building programmes. However, this can be a very time consuming process, if multiple interviews are scheduled. Closed meetings can also be treated with suspicion, especially by those stakeholders, which do not trust the independence of the process.

Surveys

Face to face surveys, conducted in a 'random' manner, can gather information from people who would not, or could not, become involved in other ways. As with all data gathering it is essential to have a clear purpose in asking the questions, and a use for the results achieved. This is an expensive method, but confidential surveys can elicit more candid results.

Response Sheets

As mentioned earlier, leaflets etc. can have reply paid cards attached, and these can be used to gather information on public concerns and preferences. These would generally have options for being included on a mailing list. There is, however, no statistical basis to the results, as the method is only as good as the distribution, presentation and in some cases the original mailing list. Results can be easily skewed.

Random Postal/Telephone Surveys

Research of this nature needs professional planning in order to attain statistically useful results, and can be labour intensive and expensive. The response rate to telephone surveys is much higher than to mail shots, but they are also more expensive to operate. That said, these surveys would generally reach a much wider cross section of the public, including those unlikely to attend meetings. The statistically tested results produced would also be more persuasive with all the stakeholders. These surveys are generally used for attitudinal research.

Internet Surveys

Precision in setting up a site for information exchanges is essential, otherwise chat rooms can generate far more input than can be looked at. The response rate is generally high but the user profile is limited by the 'digital divide,' as mentioned earlier, and so access is limited. There is of course no geographic control over the reach of the poll, it can be easily skewed and is very labour intensive.

Deliberative Polling

To measure informal opinion, randomly selected members of the public are brought together to listen to expert opinions and their own views on an issue. They are then confidentially questioned about their opinions. The result provides a representation of the publics considered judgment. This can take 2-3 days, involves up to several hundred people, requires the use of trained facilitators, and as a result can be very costly. Television cameras can be used to document the proceedings if the issue is important enough, and a resulting documentary used to air the debate more widely.

Teleconferencing

This enables face to face questions and answer sessions. This is useful where the stakeholders find it difficult to travel. This can be an expensive process, but also an inclusive one. Access to teleconferencing systems, may however be limited.

Presentations to Community Groups

At their regular meetings, social and civic clubs often invite guest speakers. The key here is to keep it short and simple (KISS), use visual aides and any other props that enable a lively presentation. The audience here may not have attended other functions, and this enables an expansion of mailing lists. The presenter has control of the information, but feedback can also be elicited. As the guest, the presenter has no input into who is invited to attend, and it is possible that project stakeholders may not be present..The audience may, however, become stakeholders.

Care must be taken to ensure the presentation is not too technical for the audience. Once prepared, the presentation can be used many times in the same format, adapting as necessary. Expert Panels. Members of local and national media are invited to publicly interview an 'expert' panel, followed by public participation. Efforts should be made to involve the stakeholders in choosing a neutral moderator, and setting the ground rules for the meeting.

This format encourages media education and gives the opportunity to dispel scientific misinformation, alongside a balanced discussion of the key issues. It does, however, require substantial organization, and may even enhance public concerns by increasing the visibility of the issues in an uncontrolled debate. The experience of the author is that, like public meetings, these are not too productive for the end game of environmental protection.

Field Offices

Where the stakeholder constituency is over a wide geographic area, the use of shop fronts in various accessible and much frequented locations, on stalls in enclosed shopping precincts and at festivals etc. can provide for a broader public outreach.

The provision of well trained staff, brochures, video and internet access points will all add to the effectiveness. Such traveling units can be brought to schools, third level institutions etc. This is a relatively expensive method, but provides opportunities for positive media coverage as well as providing for responsive, ongoing and focused communication.

Informal Small Group Meetings

These would usually be in a person's home within the neighbourhood, and involve a polite and appreciative member of staff. A coffee morning gives the relaxed setting conducive to effective dialogue. Clearly the numerical outreach is small, but the meeting being held on the communities own 'turf' and at their invitation creates a new relationship of equals.

Level 3. Involvement and Consultation

Workshops

These small public meetings (9-24 people) are designed to involve the participants in an active way. Wates, N. (2000) shows a typical format. They are a powerful tool for in-depth discussion, and with a good facilitator; the views of all those present will be heard and considered. It is also a good format for decision-making. Workshops can be used to promote involvement and discussion at larger presentations, as well, with the attendees splitting up into small discussion groups, each with a facilitator, and reporting back to the main body.

This prevents domination of proceedings, by individuals or organizations. Hostile participants, by contrast, may see it as 'divide and conquer'. Workshops humanize project team members, promote public ownership of the process, and can be targeted at particular stakeholders. Before workshops commence, it is essential to know what is to be done with the inputs received. Workshops can also be used to 'thrash out' issues raised in other formats.

Focus Groups

These are a quick means of gauging what is likely to be the public reaction to a proposal. Involving a group of 15-25 invited participants, properly facilitated, this method can give a detailed understanding of people's perspectives, values and concerns. Selection of the participants so as to involve a good cross section of stakeholders is critical. Such a forum could be a one-day event, repeated as necessary. The participants' needs for information etc. would have to be serviced. Conflicting viewpoints can be resolved, and personal respect engendered, in this as in all other small groups. If well planned and facilitated, Focus Groups are relatively inexpensive to organize.

Open House

Like any open day, this is the opportunity for the proponent, or regulator, to give a broad and detailed account of the project. The public are invited to a venue, preferably a neutral one in their own community, any time during the opening hours of 2-9.00 p.m., to enable maximum attendance. The individual is greeted at the door and the layout of information is explained. The members of the public then visit a number of 'stations', each illuminating a particular point and viewable by 6-10 people at a time. Each station is staffed by a resource-person, who can answer questions, and explain the exhibit. Visitors are asked to fill out comment sheets.

To make this a worthwhile exercise requires good publicity and the preparation of staff and materials, as well as time spent on the day. This makes it an expensive method, and one which, without added attractions, will not get much media attention. Conversely, it enables small group/one to one communication. It also allows team members to share information, and when needed, to answer visitors questions. The open house builds credibility. Open houses can be operated on the internet. Such open houses are cheap, accessible at all hours and not geographically limited. However, the 'digital divide', and the openness of the process make it difficult to use the inputs in any authoritative way, except as a source of ideas.

The Delphi Method

Following mail-shots, directed or random, with reply-paid questionnaires, the responses are collated, and then copies of the collated responses are sent to all those who responded for further comments, and so on. This enables consensus building, without the necessity for meetings and can also be done using combinations of telephone and post, or, with the caveats previously mentioned, by e-mail. It is a relatively cheap way of assessing the major issues relating to a project. The same methodology can be used to achieve expert consensus over contentious facts.

Level 4. Extended Involvement

Citizen Juries

Essentially an informal inquiry, this method involves 15-20 ordinary members of the public selected to represent a cross section of the community. They are brought together to consider a particular issue and receive evidence from expert witnesses, whom they can cross-examine. At the end of the process they produce a report setting out the views of the jury, including differences of opinion. Although its findings are non-binding, and the jury has no legal standing, the commissioning body must follow recommendations or explain why. In this way, issues can be considered in depth and in a credible way by, 'ordinary citizens'. Flaws can be pinpointed and public reaction gauged, and in a fairly short period of time. It is essential to plan the jury well, being clear, as to the resources necessary, and about what will be done with the report and its recommendations. This is a resource intensive approach, requiring the payment of jurors, a facilitator, expert's fees etc.

Advisory Groups

These can be statutory as in the case of the EPA (Section 27 of the EPA Acts, 1992) or not, and may be granted powers ranging from making recommendations to exercising leverage over final decisions. They are usually a small representative group of stakeholders who meet to discuss issues within a particular remit. Their existence is long term and not issue specific, and requires a large commitment from the participants and more resources than some of the other methods. Rules and responsibilities must be clearly defined in advance, and a consistently credible, transparent process, with third party facilitators, is essential. Potential committee members must be interviewed before appointment and the necessary support structures put in place. Participants gain a good understanding of the issues, processes and other viewpoints leading towards compromise. There is, however, the risk that the public will see the committee as too 'cosy' with the sponsor. The sponsor must be willing to take the groups advice seriously, and accept when consensus cannot be reached.

Task Forces

Generally issue-specific, a task force, made up of independent experts and stakeholders can identify, evaluate and solve particular problems. Such a group has great credibility and provides constructive opportunities for compromise. Time and labour expensive, task forces may not come to clear decisions or reach concensus.

Planning for Real

This innovative method was designed originally for communities to use in urban land use planning, but has been adopted for other situations. A good example of this latter was 'Planning for Flood Plains' (Cuff, J. 2001. p. 2.1.2.) a river basin management project. The process contained in the registered brand name 'Planning for Real' focuses on the construction of models and the use or priority cards so reducing the need for good verbal or written sills. This empowering process enables the community to identify their own aspirations and issues, bearing in mind the constraints of the real world (Wates, N. 2000. p.100). Dedicated staff time is needed and the complex 3.D models are expensive. Community Visioning. Most suited to developing general plans for a site area or organization, this has been much used by UK Local Authorities for Agenda 21 processes. This is a pre-planning process designed to collectively create imaginative shared visions for what the future could be in a community. Because there is no control over the outcome, the project proponent must be clear that considerable power has been handed to the community.

That said, this process is likely to lead to strong commitment to the outcome from all participants. Visioning can also be done via the internet, but there is no control over who makes the inputs, and restricts the process to internet users. A variation on the process is the 'Future Search Conference' used to focus on the future of an organization, a network of people, (IAP2 (b) 2003) or a community. As with most of these techniques, it is essential to use a trained facilitator. This method can involve hundreds of people simultaneously and each one is an 'expert'. A 2-3 day meeting, this may prove logistically challenging. It may also prove difficult to gain complete commitment from all stakeholders.

Evaluating the Process

Evaluation has been defined as 'a process of assessment which identifies and analyses the nature and impact of processes and programmes' (Interact, 2001). The evaluation of public participation is, however, a relatively rare activity (Gray, A.J. 2002. p.46). Yet it is an essential feedback mechanism during the process for the stakeholders, and post process for the funders, proponents, and stakeholders, to see, if the process is going well, or whether it was worth the effort. As a learning process for future processes, clearly, record keeping and assessment are essential. Clearly then, it is not just the end result that is of importance in this assessment, but the journey to reach it. Most of the international evaluations use 'fairness; and 'competency' as underlying criteria (Webler; T. 1995), and most rely on 'generic' expert criteria. But Barnes, M. (1998) suggest that the evaluation process itself has to adopt participatory techniques if the criteria for assessment are to include those considered important by all the stakeholders. Looking at the assessment of participatory processes for the UK Environment Agency, Petts. J. (2000) uses some 12 effectiveness criteria.

- o Clarity of objectives-is the connection made between purpose process and outcome?

- o Clarity of legal process-including limitations of outcome.
- o Consensus on Agenda and Procedures including choice of facilitator or chair person.
- o Input into assessment process.
- o Representativeness.
- o Inclusivity-are all barriers to participation minimized, and are non-participant stakeholders kept informed?
- o Deliberation-is dialogue open, inclusive, informal, detailed enough, educational and constructive?
- o Capability-are sufficient resources in time, money and staff training available?
- o Sound learning-is the process enjoyable and informative? Does it lead to increased consensus?
- o Actual Decision Responsiveness-is the process outcome included in the final decision made by the relevant authority?
- o Trust Enhancement-has the process increased trust between proponent and stakeholder for future participatory processes?

Perhaps a more useful tool, however, is that developed for the EU Life-Environment Project, 'Wise use of Flood Plains' (Cuff,J.,2001,pps 75-81). This could be used for an individual public participation method or for a whole public participation process. Figure 2. Cuff's Circle. Summary framework for evaluation of participatory processes. (Cuff, J., 2001) The categories are adapted here, where necessary, for a more general use by the author. The circular process, with 'outcomes' as its centre, and about which all else revolves, is a useful visual to maintain focus throughout the process. Objectives-These could be the objectives of the proponent and/or the objectives created as part of planning for the process. The two sets of objectives could be compared to see if participation objectives complement strategic ones. Did the objectives change over time and were they realistic. To what extent were objectives met. Contexts-Is the process connected to: for example rule making, long term planning, licensing, land use planning or a combination of two or more, e.g. a planning application with an EIS and an IPPC Licence process all on the one project? If so, how did the different processes interlink?

Other factors could be political (e.g. local elections) or geographical (e.g. a long powerline where jobs depend on it at one end, but people along its route resent it). Increased pressures on a process can be due to a need to act, brought about by external factors, e.g. flooding of the Somerset Levels (Cuff, J. 2001. p.77). Levels of Involvement-Using Arnstein's ladder, methodology can be assessed and explanations given for the level of power transfer used, in the light of the submissions made. Inclusivity and Stakeholder Identification-This should also be measured in the light of the objections. Who was left out, or included? Were there complaints about this? Was a special effort made to include socially excluded groups? Were events timed to suit stakeholders? Was there a reluctance to participate? if so, why? How many, and what type of people actually took part?

A format. that could be used to evaluate individual participatory methods or processes. Scoring: 0= went badly, did not meet objectives, 10= all objectives met, total success. Participatory Methods/Processes-This assessment should be for each method used, and an objective set for each at its onset. The use of an H Diagram gives a good picture of the pros and cons of an event. All aspects of the methodology should be scored from 1-10 to show strengths and weaknesses, and recommendations for the future use of same. Innovation-Any new methods could be shown and assessed here, for possible use in the future.

Commitment to using Participation Process Results-Was it clear at the outset what limitations the participation process had in influencing the decision-making? What actual use was made of the process outcomes? Were participants informed of the use of their inputs? If so, when and how? Inputs-A break down of the costs of the process into individual events, and then into their constituent costs, is needed to assess the value-for-money represented by the different methods. Was lack of finance an obstacle? Was the use of consultants value for money? Outputs-This is a measure of the number and quality of outputs and their effectiveness, e.g. posters, exhibitions, questionnaires, etc. Outcomes-These stem mainly from 5 and 9 (figure 2, above) and are measured against the baseline of the pre-process situation. Firstly, there are the strategic outcomes. Has a decision/recommendation been made, and is it based on a consensus? Secondly, there are the cultural outcomes. For stakeholders these include:

- o Change in the attitude to participatory processes.
- o Ownership of the process o Increased trust in the proponent
- o Changes in values or attitudes

For organizations:

- o Are they more transparent?
- o Are they more at ease with public participation?
- o Do they see the process as worthwhile?

There will be other outcomes in the many circumstances in which this assessment process could be used.Wates, N. (2000, p.100) describes how to organize and run an effective 'review session' involving all stakeholders.

Conflict Resolution

Having pursued a participatory programme what can be done if the stakeholders cannot reach concensus? The use of strategies known as ADR (Alternative Dispute Resolution) and based on 'joint gains' problem solving, mediations, facilitation and consensus building offer promise for many cases (Adler, P.S. et al., 2003). Bingham, G. (1986) showed that 78% of the 161 environmental disputes studies, which used mediation-based ADR, ended in agreement amongst the parties. It is, however, not a panacea, and each case should be judged on its merits. The US EPA uses the definition of ADR in the Administration Dispute Resolutions Act of 1996 "any procedure that is used to resolve issues in controversy, including but not limited to conciliation, facilitation, mediation, fact finding, minitrials, arbitration and the use of ombuds or any combination thereof. All the methods listed above involve a neutral third party who assists others in designing and conducting a process aimed at reaching agreement. All aspects of ADR are voluntary, including the decision to participate, the type of process used, and the content of any final agreement. The ADR process and practices are beyond the scope of this study and generally would only be undertaken using specialist EDM mediators or facilitators, however, it is worth looking at the underlying principle. The conditions, under which ADR is useful, and most likely to be successful are listed by Otolano (1997. p.417) as:

- o A high uncertainity in timing and outcome. Stakeholders who perceive other quicker or more certain routes to achieve their objectives all have little motivation to become involved in ADR.
- o Equivalent distributions of power. Here power is described as the ability to prohibit an action desired by one or more adversaries.
- o Mediation is more promising than BATNA (Best Alternative to a Negotiated Agreement). As long as the prospective outcome from negotiation is more attractive than BATNA, a stakeholder has an incentive to continue negotiating. If the prospects for success for negotiation drop than the BATNA might be the courts, lobbying or direct action.
- o Disputing parties are at an impasse.
- o Litigation is ineffective.

Ortalano, L. (1997, p.411) goes on to describe the ADR techniques known as 'principled negotiations'. This is characterized by four main points, which are a useful insight into problem solving during any EDM process.

- o Separate the people from the problem
- o Focus on interests not positions
- o Invent options for mutual gains
- o Use objective criteria for evaluating the agreement.

Implicit in these is the recognition of the four main types of conflict delineated by Creighton, J.L. (1981. pp 454-455).

- o Cognitive conflict-differences over the facts.
- o Values conflicts-a dispute over goals, for example a dispute over whether a particular outcome is desirable (undesirable) or should occur or not.
- o Interest conflicts-Cost and benefits are rarely evenly distributed so that some may have a greater interest in a particular outcome. It is possible then to have agreement on facts and on values but still disagree on interest.
- o Relationship conflict-Communication involves both facts and relationships (how much a person is valued, accepted etc.). The EDM process itself may favour those with power and money and so communicate a lesser value to other stakeholders, leading to conflict on emotionally motivated grounds other than facts, values or interests.

CAPACITY BUILDING

Defined by Wates, N. (2000) as "the development of awareness, knowledge, skills and operational capability in order to achieve a purpose", capacity building is an essential part of the development of public participation practice, and it is necessary to build capacity amongst all the stakeholder groups in society.

Agenda 21, Chapter 36 set out the broad proposals for promoting education, public awareness and training which it sees as meeting capacity building needs. It promotes the inclusion of environmental and human ecological considerations in all aspects of education and decisionmaking.

'The Access Initiative' (Coyle, J. et al. 2002) in a study of nine nations found that: lack of government capacity constrains public access e.g. through lack of staff, equipment, procedures and training. Laws and procedures relating to access are often new, and the necessary training has not been done to facilitate their enactment. South Africa was seen as the only country, of the nine studied, with an extensive programme building staff capacity. Lack of capacity in civil society was also seen as a serious impediment in all the countries studied.

The Environmental Law Institute (Breggin, L. and Hallman, H. 1999) looked in detail at Building Capacity to participate in US EPA activities and pinpointed the following as essential elements in capacity building.

- o Information.
- o Technical Assistance.
- o Process Education.
- o Access to Documents.
- o Education on laws as fundamental building blocks.

The study then went on to give potential approaches to enabling stakeholders to participate, showing models of each method in action. These include;

- Independent Information Broker
- Ombudsperson
- Hotlines
- Technical Assistance Grants
- Citizen Training on Processes and Legal Requirements
- New Collobrative Participation Processes
- Increased Date Availability and Dissemination Networks
- Grants to Community Groups
- Improved Access to Documents
- Improved Mailing Lists

Capacity building in the Irish context will be looked at in the following chapters, but the following are a few examples of best practice at work elsewhere. The US EPA's 'Community Based Environmental Protection' programme is another way to create local ownership of the environment (US. E.P.A., 1999). The many Local Agenda 21 projects worldwide are achieving similar results. In the UK, through the Environment Council, training programmes for all stakeholder groups are available on stakeholder dialogue, as a tool to achieve sustainability. Information tools to enable local government employees to respond to their publics now include a one-stop internet portal. In Ireland, IBEC runs regular training programmes for people in Industry, on 'Environmental Communication', to promote better communication with the other stakeholders, regulatory bodies, the media, and the local community (IBEC, 2003).

IRELAND AND PRINCIPLE

The Access Initiative

Over the last three years the World Resources Institute has formulated a methodology to assess the implementation of Principle 10 at national level, called The Access Initiative (TAI) (The Access Initiative, 2003). This is a global coalition of civil society groups collaborating to assess and promote national implementation of Principle 10 commitments. The European Commission has now signed up to Partnership for Principle 10 (P.P.10), which commits governments to support such independent assessments, by NGO's, of national Principle 10 performance. The pilot project, reported in 'Closing the Gap' (Coyle, J. et al. 2002) involved nine nations on five continents and twenty-five civil society groups. Following this, the process was refined, and the difficult task begun of finding a methodology for assessing the implementation of Principe 10 in the many different cultural and political systems that exist. The methodology looks at:

(i) The comprehensiveness and quality of the General Legal Framework

(ii) The degree of available access to selected types of information about the environment.

(iii) The degree of public participation in decision making processes, and

(iv) Comprehensiveness and quality of capacity building efforts to encourage informal and meaningful public participation.

In order to conduct an assessment of the Irish government's performance, and so to pinpoint weaknesses, with a view to capacity building, the Access Initiative requires the building of a coalition of civil society groups within Ireland.

The formation of a coalition is necessary to get the skills and knowledge base necessary to perform the research and analysis tasks contained in a full Access Initiative Assessment. Potential coalition members might include NGO's, lawyers, academics, journalists, etc. The structure of the coalition would include a local organization, a research team with membership from across the coalition, and an advisory group of 'experts' to oversee the research and give credibility to its findings (The Access Initiative, 2003).

The work involved, with 6-12 researchers, is estimated to take 4-6 months, based on the pilot project. Clearly the degree of detailed research required to carry out a complete TAI study of Ireland is beyond the scope of one individual and of this study, but a selective look at the issues raised will give some insight into the current state of environmental governance in the Republic.

Access to Information

Law: Whilst there is not constitutional right to information on the environment, freedom of access to information is available under the Freedom of Information Act, 1997 (FOIA) and the European Directive 90/313/EC (S.I. 125 of1998) (replaced by 2003/4/EC). Access under FOIA is restricted by cost, limiting its use by people with low incomes. Emergencies: Although a general legal mandate exists (The Seveso Directive) for dissemination of information about environmental and health impacts to the public during and after an emergency, An Taisce has repeatedly expressed its concern about the implementation of same (Lumley, I. 2003)

Monitoring: In general, monitoring of air and water quality is carried out to a high standard and information is made available about the results of same. Both air and water quality monitoring results are available on the EPA website, and are freely available to the public through ENFO. The quality of this information is of a good standard, but the data is 1-2 years old. Local Authorities also carry information on their web sites, though the range and quality varies considerably.

State of the Environment Report (SOE): The EPA, as mandated by the EPA Act 1992, produced its third SOE Report in 2000, and it is rich in data and indicators to describe conditions and trends. The report was well publicized in the media, and available in bookshops. 'Government in Focus-Key Environmental Indicators For Ireland 2002', is available on the web-site. Facility-level Information: Readily available and timely information on the environmental impact of particular facilities is essential to enable people to protect their environment. In the USA online information regularly updated is available on the Toxic Releases Inventory (TRI) web-site. In Ireland the EPA is developing a similar system, based on GIS mapping. Local Authorities have some information on web sites, but not facility specific. Otherwise it is necessary to trawl through registers at EPA or Local Authority offices.

Participation in Environmental Decision Making

Law: The legislative and constitutional frameworks regarding participation are beyond the scope of this study, but would be addressed in an Access Initiative Assessment. Whilst public participation is allowed at an early stage in some environmental decision making processes (e.g. Development Plans) in others it happens too late to make substantive changes (e.g. the EIA process), and in others, there is no right of appeal to an independent arbiter such as An Bord Pleanala (e.g. Article 81,Planning and Development Regulations, 2001). Policy: Lead times for notification of public participation are acceptable in terms of the Access Initiative, though the author's experience would suggest that this depends on the effectiveness of the notification procedure. There is a wide availability through ENFO, the Internet, Local Government Offices, Libraries, etc. of policies, strategies, plans programme and laws.

The requirement to send development plans, planning applications etc. to the 'statutory bodies' gives a degree of external consultation by non-involved expertise, but as explained elsewhere, much environmental decision making in Ireland is of the 'report and comment' type. Project and Plans: The lead in time for public comment on draft plans is generally eight weeks, but it is five weeks for a planning application, regardless of the project size and complexity. The quality of information provided to the public and relevant authorities, particularly EIS's is very varied, and is often, in the author's experience, poor both in content and presentation. Clearly this does not allow for good decision-making. Records of local permits and other project documents are accessible at the relevant offices and increasingly on the internet.

Capacity

Law: The right of freedom of association and the right to a clean environment are guaranteed under the Irish constitution.

Government: The establishment of FOI officers in all the government departments dealing with environmental issues, and the publication of the EPA Section 15 and Section 16 reference books are some of the ways in which capacity is being built by the government with regard to access.

The presence of ENFO (The Environmental Information Office) as a one-stop shop and library for all environmental information is a very important capacity building force. ENFO also provides support for schools and materials for environmental education. ENGO's are encouraged and give some funding under European Parliament Decision No. 466/2002/EC. It is not intended to give further comment here on access to EDM in Ireland as this will be discussed later in the light of the results of the survey undertaken as part of this study.

ENVIRONMENTAL IMPACT ASSESSMENT (EIA) AND PUBLIC INVOLVEMENT

EIA is a gateway control and not in itself a permitting or authorisation process. It is a very important tool to enable good decision-making, and although its findings usually correlate to the final decision, theoretically the reverse can be true. Article 6 para 2(e) of the Aarhus Convention states that "the fact that the activity is subject to a national or trans boundary impact assessment procedure", dictates that the public concerned shall be informed, either by public notice or individually as appropriate, early in an adequate, timely and effective manner".

Further, Articles 4 and 5 require each Party to provide for public participation when all options are open, whilst encouraging applicants to identify the public concerned and enter into discussions prior to applying for a permit.

Within Ireland Council Directive 85/337/EEC as amended by Directive 97/11/EC lays down the requirements for the EIA process, and is given effect in Irish Law through the Planning and Development Act 2000 (implemented by The Planning And Development Act, Regulations, 2001.), through IPPC Licensing and Waste Licensing and through the European Communities (Environmental Impact Assessment) Regulations, 1989-1999. (EIA Regulations) (Doyle, A., 2002).

In the EIA Regulations, a developer when required must submit an Environmental Impact Statement to the competent Authority. An EIS is defined in S.I No. 349 of 89 (Article 3 (1) as follows " A statement of the effects, if any, which a proposed development, if carried out, would have on the environment". Following the submission of an EIS, the public, relevant public bodies and, where legislation requires, other governments, have the right to comment.

The competent authority must take all submissions into consideration and make a decision. The decision and the reasons for taking it must then be communicated to the public and other interested parties. The whole of this procedure is the EIA. There is no publicly available written analysis of the process leading up to the decision, unlike the IPPC and Waste Licensing processes. In these, an inspector's report to the Agency, containing an analysis of the issues, is available for public comment, in the course of an objection. Essentially through, in each of these three processes it is the presentation of a report and then public comment.

The EPA by contrast in its "Guidelines on the Information to be Contained in Environmental Impact Statements"(P.7), lists public participation as one of its four basic tenets of EIA. In the same guide (para 2.3.7) it suggests, that apart from its basic statutory rights, the public could participate in the EIA process by making observations or submissions, in response to a preliminary invitation to participate in the procedural stages of scoping; evaluation of alternatives, and document review. Public involvement early in the process enables applicants to include or address concerns before it is too late.

It further asserts that, direct participation, i.e. face to face, communication is more likely to result in accurate and focused communication. In Para 2.3.6 the role of NGO's is discussed, and their input encouraged at the earliest stage possible. "Early, open and constructive engagement has frequently proven to be beneficial to both the protection of the environment and to the quality of development projects", it states. However, at present these are recommendations only, and it is the proponents decision whether or not to involve the public, apart from the review of the final EIS.

The EIA Process

The Definition: "EIA can be defined as: the process of identifying, protecting evaluating and mitigating the biophysical, social and other relevant effects of development proposals, prior to major decisions being taken and commitments made" (Saddler, B. et al, 1999) The World Bank expands this definition and describes EIA as a procedure that; "evaluates a projects potential environmental risks and impacts in its area of influence; examines project alternatives; identifies ways of improving project selection, siting, planning, design, and implementation, by preventing, minimizing, mitigating or compensating for adverse environmental impacts and enhancing positive impacts" (Klees, R. 2002).

The main aim of the process is to stimulate thinking, and encourage action, and not just the ticking off of boxes just to get another report (Verheem, R. 2002).

The Principles of EIA

The IEA (Saddler, B. et al., 1999) identifies two sets of principles, 'basic' and 'operative'. The basic set applies to all stages of EIA as well as to Strategic Environment Assessment (SEA) of policies, plans and programmes. One of these states that the EIA process should be participative; providing appropriate opportunities to inform and involve the interested and affected public, and, that their inputs and concerns should be explicitly addressed in the documentation and decisionmaking" (Saddler, B. et al., 1999)

In a review of 25 years of National Environmental Policy Act (NEPA) the Council on Environmental Quality (CEQ) (CEQ 1997) stated "the success of a NEPA process heavily depends on whether an agency has systematically reached out to those who will be most affected by a proposal, gathered information and ideas from them, and responded to the input by modifying the proposal or adding alternatives, through the entire course of the planning process".

The EIA Cycle and Public Involvement

The EIA process should be applied as early as possible in decision making for a project, providing for the involvement and input of communities and industries affected, as well as the interested public (Saddler, B. et al., 1999)

The EIA cycle shown in Fig. 4 (above) as part of the planning system and project cycle should provide for the following: (European Commission, 2000).

Screening

Screening determines whether or not a proposal should be subject to EIA, and if so, at what level of detail. Consultation with potentially effected parties here will improve understanding of the nature and significance of potential impacts using 'local' knowledge, together with that of 'experts'.

The objectives here should be: to obtain a complete understanding of how the issue is viewed by all the significant stakeholders; and to identify the future levels of interest in public participation activities on this issue. Techniques could be used here from levels 1 and 2 of the participation tools list. (Eccleston, C.H. 2000. p.65) has a checklist of tasks to consider as part of the prescoping phase. Added to this should be the preparation of a public participation plan, which includes a publicly available record of the process. The use of exploratory thinking techniques such as brainstorming and snow-storming at this phase helps stimulate the creativity need to seek alternatives.

Scoping

This is a very early exercise in an EIA in which an attempt is made to identify the attributes of the components of the environment for which there is public (and professional) concerns and upon which the EIA should be focused (Singleton, R., Castle, P. and Short, D. 1999, p.103). Good scoping is the key to a successful EIA process. Public involvement here ensures that all the significant issues are identified, local knowledge about the area is incorporated, and alternatives are identified and considered. The principal objectives of public scoping are: (Eccleston, C.H. 2000, p.72)

- Identify public concerns and the expertise needed to investigate same.
- Identify alternatives to be examined
- Identify significant issues that need to be analysed, eliminating the unimportant.
- Identify problems and potential solution early in the process.
- Identify problems with the participation process and address same.
- Ensure that both the positive and negative aspects of the proposal are identified and studied.
- Identify potential mitigation measures.

A scoping information package should be put together to promote public involvement and to inform the scoping delegates. It should include:

- An invitation to participate showing how, when and where.
- A brief description of the working of the EIS process and the opportunities for participation.
- A description of the proposal, such that the objectives of the proponent are clearly laid out with maps, diagrams, figures etc.
- A description of the known potential impacts. A public notice should then be issued using any of the methods deemed necessary from Level 1 (Section 1).

As was mentioned in the section on P.P. tools, public meetings, whilst commonly used are not the best method, unless they are based on small group sessions and workshops. Public opinion surveys, citizen advisory committees or any other methods listed in Section No. 2 Simple methods may suffice for describing, synthenisizing and communicating information on the preproject environment and the potential impacts e.g. using checklists, matrices and networks. (Eccleston, C.H. 2000. p.75) (Australian EIA Network, 1996) gives a check list for planning a public scoping meeting, and warns that "outside entities that participate in scoping generally do so because they are opposed to the proposal; the remaining participants who support the proposal often do so because they stand to gain from it. Not surprisingly numerous public scoping efforts have dissolved into sessions of frustration, dissension or outright confrontation. So it is essential to plan well and use professional facilitators or neutral moderators where debates are likely to be heated.

Alongside the general public scoping, it can be of value to conduct focus group/workshop meetings with stakeholders having special interests or expertise, in order to examine more detailed or complex issues. The results of these meetings should be made public, and so become part of the general scoping process.

Eccleston, C.H. (2000 p.249) gives a methodology for handling the potentially voluminous, or complex scoping input, followed by an internal 'responsible authority' scoping session. This can consider issues including technical ones, using in-house expertise, and in greater depth than might otherwise be possible in a public setting.

The body conducting the EIA process, together with stakeholder representatives should then produce a post scoping document, summarising the scoping process, its findings and the reasons for decisions taken in reaching the scope for the resultant EIS Implementation Plans.

Impact Assessment/Mitigation

"Public involvement here can serve to ensure that the analysis and mitigation, necessary to avoid, minimize or offset predicted adverse impacts is relevant to local concerns, and accurately reflects local value and preferences". (European Commission, 2000).

Both impact mitigation, and the evaluation of significance are complex and time consuming processes, which cannot be dealt with by large groups. It is necessary, therefore, to use stakeholder representatives to work with the experts, reviewing the finding of the experts in workshop sessions or other facilitated small groups. Regular reporting of progress in the process, to the wider publics will maintain their 'ownership' and sense of involvement.

The Evaluation of Significance

This is a difficult issue to resolve even between specialists. It is, however, essential that the public be involved and their perspective included, as the interpretation of significance occupies a fluid boundary between science and politics (Australian EIA Network 1996).

The Canadian EA system of impact assessment is based solely on scientific, credible technical and other relevant information (C.E.A.A., 2001. The resulting determination of significance must be 'objective' and reasonable so as to withstand court challenge. By comparison in the US system, under NEPA, public opinion and the controversiality of the proposal help to identify and determine significance. Ultimately, however, the relevant authority will decide, taking all inputs on board.

Evaluation of significance is subjective, contingent on values and dependent on the environmental and community context (Australian EIA Network, 1996). The intrusion of wider public concerns and social values, into the significances evaluated by scientists is inevitable, as discussed elsewhere. The challenging nature of this part of the process is, therefore, one in which the use of facilitated small groups would be essential for the more contentious issues. More information, on the technical aspects of the evaluation of impacts. is given by Lein, K.L. (2003).

Preparation of EIS

It is essential to document clearly and impartially the impacts of the proposal, as well as those of the identified alternatives, the proposed mitigation methods, the significant effects and the concerns expressed by the public and communities affected by the proposal, as well as how those concerns were addressed.

The draft EIS, following internal review by the relevant body, should be made available for public review, and should include a non-technical summary. Review of the EIS The review of the EIS should determine whether the report; meets its terms of reference, provides a satisfactory assessment of the proposal(s); and contains the information required for decision making. Involvement of the public can ensure the quality and comprehensiveness of the assessment and help to reduce any bias in the analysis.

Before the public review, the EIS should be complete in every aspect, except this final opportunity for public feedback. Any major changes resulting from this public review would probably trigger a second public review, though good public participation early on should make this unlikely.

Proposals for electronic transmission of the document along with public access through libraries, site offices, public displays, information repositories, and presentations, should give wide availability. Methods of feedback should be widely publicised. One method could be an interactive web-site page for this purpose.

All those who made major contributions to the process should receive a copy (finance permitting) or have access to a copy, along with the statutory bodies and the proponent. The draft should be available and publicised at least 15 days before any stakeholder meetings to discuss it.

The final EIS should then include the changes based on the relevant feedback in the draft, together with the responsible body's reasons for not including any other issues raised. Eccleston, C.H. (2000) gives useful guidelines on dealing with this stage of the process.

Decision Making

It is generally the role of the regulatory authority to approve or reject the proposal and, if it is approved, to establish the terms and conditions of its implementation, taking into account the EIS, and its public input.

Follow-up, Implementation and Monitoring

The participation of local representatives and NGO's in monitoring the operational impacts of a project can lead to the early identification of problems, and can foster a sense of public partnership.

Continual assessment of the EIA process

Continuous assessment of the process itself can ensure the implementation of the public participation plan, provide public transparency and strengthen the effectiveness of future EIA's.

EIA Best Practice

In its 'Directory of Impact Assessment Guidelines' (Donnelly, A., Dalal-Clayton, B., Hughes, R., 1998) the IIED (International Institute for Environment and Development) concludes that "greater attention to stakeholder involvement leads to better environmental assessment, and thus to the formulation of projects that deliver more social benefits, fewer environmental costs and greater economic and financial benefits". In looking at best practice world wide, it will not be possible to make assessments of their effectiveness within the scope of this study. However, it finds that in reality EIA guidelines and practice generally pay only lip service to real stakeholder involvement.

This is confirmed by the European and Central Asia Environmentally and Socially Sustainable Development Department (ECSSD) regional study of Eastern European and Central Asian Countries (Klees, R. 2002) where quite often, inclusive legislation is present but the reality of implementation is quite different. Constructive stakeholder involvement in EIA will rarely occur spontaneously. A proactive approach is required; starting with clear guidelines, legally enforceable by inexpensive court action; and financial assistance for stakeholders where significant costs would be incurred in order to take part in dialogue.

There are a number of good examples of proactive legislations and stakeholder support systems. Canada. Public participation in the federal assessment process is an implicit part of the Canadian Environmental Assessment Act, 1995, which was itself the result of five years nationwide consultation. The Canadian Environmental Assessment Agency (CEAA) was established to administer the Act and to ensure that opportunities for public participation are provided (C.E.A.A., 2001). The CEAA must reach out to the public for their input into screenings, comprehensive studies, mediations and panel reviews.

A public registry is established to ensure public access to records, and the responsible authority must address the need for a follow up programme to verify the accuracy of the assessment and efficiency of the mitigation measures.

A participant-funding programme, designed to provide limited funding to interested individuals and groups, is open both to those for and against a project. Its purpose is to enable effective participation at key stages of mediations and panel review.

The 'Guidelines on Environmental Assessment and Traditional Knowledge of Indigenous Peoples' (Donnelly, A. et al., 1998. p.29), formulated in Canada, draws attention to the limitations of short term scientific procedures to collect reliable and adequate information on which to base predictions. They also explore the very different perceptions of environment and development held by indigenous and non-indigenous stakeholders. New Zealand.

The Resource Management Act 1991 (RMA) states the two key principles: that decisions on environmental matters are most appropriately made by the communities most affected by those decisions, and that community participation is vital to effective resource management. This legislation is based in part on the Maori 'Kaitia Kitanga' (guardianship) of the 'Mauri' (life essence binding the physical and spiritual elements of all life). It is the first attempt world wide at an holistic legislative approach to the management of environmental resources (Ministry for the Environment, New Zealand, 2003).

The Act simplifies access for the public by putting all activities that impinge on environmental resources under the same resource consent procedures. A resource consent gives a person or organization permission to sue or develop a natural or physical resource, and/or carry out an activity that affects the environment in some way for a stated period. Resource consent replace all the many different permissions granted under previous law. Its simple appeals structure, Environmental Legal Assistance Fund, and its insistence that decision makers actively consider alternatives to regulation in preparing policy statements and plans (Section 32 of the RMA), makes the RMA a highly accessible process for the public to become involved with. The Arctic.

The Arctic Environmental Protection Strategy (AEPS) agreed by the eight Arctic countries lays great importance on the integration of two ways of knowing: Traditional Indigenous Knowledge and Scientific Knowledge (A.E.P.S., 2002). Experience gained in mapping traditional knowledge on the Belluga whale in ten indigenous communities on the Alaska and Russian sides of the Bering Sea, have lead to the inclusion of the use of traditional knowledge alongside general public participation as two of the important features of the "Guidelines for EIS in the Artic"(A.E.P.S., 2002)

The USA. The Council on Environmental Quality (CEQ) has been central in strengthening procedures and guidelines in the USA since the National Environmental Protection Act 1969 (NEPA) was enacted. Public scoping of all federal agency projects requiring an EIA is the norm. Publicity in various media is combined with invitations for comments or suggestions, which can be submitted via the post, at various public open houses or public meetings, and via e-mail. Sri Lanka. The scoping process here involves formal and informal meetings with people who may be affected by the proposed project either directly or indirectly, or who may have special knowledge of the project area and its environment (Tille Kerante, L.S.G., 2003). Australia. The Environment Protection and Bio Diversity Conservation Act 1999 (EPBC Act) establishes a Commonwealth process for environmental assessment and approval of proposed actions likely to have a significant impact. This is the culmination of over 25 years of EIA in Australia.

Five options are listed under Section 8 of the EPBC Act (Environment Australia, 2003). The first three, preliminary documentation, public environment report (PER) and environmental impact statement involve the following steps: proponent prepares and publishes draft document; public comment period; proponent finalises document incorporating comments; Secretary of Environment Australia reports to the Commonwealth Environment Minister who makes a decision. All documents prepared by the proponent will follow guidelines (terms of reference) laid down by Environment Australia, and formulated by consultation with the relevant stakeholders. Where the impacts are likely to be high, and prior to preparation of the final EIS, or where greater public involvement is to be ensured, a Public Inquiry is held. All information leading up to the inquiry, including the reasons for having same, are made available. The hearings are all held in public and all submissions made available to the wider public. Quebec, Canada. (Australia EIA Network, 1996 p.25) Established in 1978, the Bureau d'Audiences Publiques Sur L'Environnement (BAPE) is a quasi-judicial advisory body with a strong record of public participation. The Minister on request from any person, group or municipality may instruct BAPE to hold; a fact finding enquiry; carry out a combined enquiry and mediation; or undertake public hearings. Its role is to enquire into any question, relating to the quality of the environment, that has been submitted to it by the Minister of the Environment and report back on its findings. It is not like An Bord Pleanala, which is an appeals board only. B.A.P.E. once instructed, plays a proactive role.

A BAPE Commissioner meets with the parties and mediates to seek a joint solution. Failing this, two to five Commissioners serve as an independent review board. Initially, the public has 45 days to review an EIA and request mediation or a hearing. BAPE must report to the Minister, within 120 days of receiving a request for a public hearing or 60 days from receiving a request for an inquiry or inquiry with mediation. Success rates of mediation by BAPE are 75%. The option for a further period for a hearing helps circumvent the feeling of negotiating under time pressure. At the same time, proponents know that these timescales are absolute. The mediation process provides an effective and efficient means of resolving conflict with a 75% success rate in 1978. Institutional arrangements for EIA Review (Australian EIA Network (1996). Permanent Commissioners in Holland (Netherlands Commission for EIA, 2002) and Italy review the quality of the EIS. The Dutch Commission for EIA remains outside political decision making, it's role is to give an expert (selecting from a panel of 400) opinion on the quality and correctness of an EIS. The Italian Commission by contrast can make a judgment on the environmental compatibility of a project, as well as the quality of its EIS. In Canada, for certain major projects (50 in the period 1975-1995), ad hoc review panels, or inquiry bodies are used. A panel's major responsibility is to establish terms of reference (TORs) for an EIS and to conduct a public review of its suitability. The Parliamentary Commissioner on the Environment in New Zealand can also use panel reviews.

Ireland, Planning Major Roads. Although the Roads Act, 1993 as amended by the Planning and Development Act, 2000 only specify public consultations in the case of compulsory purchase orders (CPO's) or post EIS publication, the NRA in its 'National Roads Project Management Guidelines' sees public consultation as a vitally important and integral part of the various phases of major project planning. The first round of public consultation occurs in the preliminary planning stages as part of the feasibility/constraints study. Following the publication of the Consultants Report, which is based on the public submission and the consultants report, the possible route options and their general impacts are identified and the public consulted again. This leads to the route selection, and only then, is the EIS drafted. (NRA, 2000).

Research Into EIA. The EC Joint Research Centre (ISPRA) is at present conducting a study on behalf of the EC Environment DG into EIA and SEA research. One of the six areas for research relating to EIA's is public participation (Europa, 2003). A more general research 'Programme on Environmental Decision Making' (2001-2006) (E.S.R.C. 2001) aims to investigate and elucidate a set of criteria for better environmental decision. Further examples of good practice in public involvement in EIA and other areas of Environmental governance can be accessed through the good practices suite of the United Nations website: Virtual Conference for Integrating Environmental Considerations into Economic Policy Making Processes (ESCAP).

Strategic Environmental Assessment (SEA)

"SEA is a systematic process for evaluating the environmental consequences of proposed policy, plan or programme initiatives, in order to ensure they are fully included and appropriately addressed at the earliest appropriate stage of decision making on part with economic and social considerations". (Sadler and Verheem 1996). In their principles for SEA, Dalal-Clayton and Sadlier(Donnelly, A. et al, 1998, pp 31-42) call for transparency and openness, including early consultation, availability of comprehensible documents and explanation of how all decisions are arrived at.

The development of a pro-active SEA process is a subject too big for this study, but is one in which public participation should play a part as large as that proposed for the EIA process.

The Dutch systems for EIA and SEA are identical in structure, and in both public involvement at the scoping phase is required. (Netherlands Commission for EIA 2002). By contrast, the new European Directive leaves consultation until much later in the process. 'Directive 2001/42/EC (The SEA Directive) (117) states that prior to the adoption of a plan or programme, or its submission to the legislative process, the competent authority must carry out an environmental report.

The draft plan or programme and the environmental report must be made available to the relevant authorities and to the public, who will be able to express their views on same, prior to it's adoption or submission to the legislative process. The relevant authority, following where necessary trans-boundary consultations, will inform all the parties concerned which have been consulted of its decision. With the decision will be a statement summarizing amongst others the results of the consultations, and the reasons for choosing the final plan or programme.

THE SURVEY

The purpose of the survey was to make an assessment of the present state of public participation in EDM in Ireland. An attempt as made to look at:

- o Attitudes of the main stakeholders.
- o Their experiences with public participation.
- o Present in-house capacity of regulators and industry to enable public participation.
- o In-house capacity building measures.
- o Capacity building in the wider community.

Methodology

The questionnaire was based on the results of the literature survey, personal experience and feedback from trial questionnaires. The survey was conducted by e-mail, except the industry section, which due to confidentiality issues had to be posted by the college administrator. Where appropriate, the same questions were asked of all stakeholders, but specific questions relevant to each group were also included.

The public questionnaires included a preliminary set of open-ended questions. These were designed to establish the person's particular experiences and involvement in dealing with a particular EDM process.

Copies of all the questionnaires can be seen in Appendix III.

The main groups of stakeholders surveyed were as follows.

The Public.

Industry.

Local Authority, Environment Sections.

Environmental Awareness Offices (EAOs).

Department of the Environment and Local Government.

Other relevant departments and agencies.

All respondents were guaranteed confidentiality.

The Public-20 completed questionnaires.

Notice of the survey was given to all the ENGOs on the ENFO address list, and information about the research was given in the 'Horizons' column of the Irish Times (21/12/02 and 22/3/03).

All the respondents had been involved in some form of EDM process, either individually or more usually in groups of from 4 to 2000 members.

The areas of EDM covered by the respondents included.

- o Urban, Rural, Waste, Landscape, Road and County Development Planning.
- o The EU Habitats Directive.
- o Integrated Coastal Zone Management (ICZM).
- o High Voltage Power lines.
- o Compliance with EC Directives.
- o Aquaculture.

o Incineration.

o Recycling.

Industry-15 Completed Questionnaires.

Postal contact was made with 40 graduates of the H.Dip in Environmental Protection. The responses came mainly from Environmental Engineers and Environmental Safety Officers, together with one consultant and one company CEO.

The industries involved included:

- Wind energy Distillation
- Pharmacies Biological
- Metallurgy Galvanizing
- Chemical Coating Turbines
- Aircraft Man-made fibres

Local Authority, Environmental Sections-12 completed questionnaires.

Questionnaires were circulated to all 34 local authorities, some of which did not have separate environment sections. A 35% return rate.

Environmental Awareness Officers (EAOs)-16 completed questionnaires. A 47% return rate.

Although based in local authorities, the function of EAOs is separate.

Department Of The Environment And Local Government-1 completed questionnaire.

This return represents both the individual opinions of the respondent and department policy where appropriate

Other Relevant Departments and Agencies-2 completed questionnaires.

One was from a Senior Environment Officer with a Fisheries Board, the other the Coordinator of a River Basin District (RBM).

Calculating Percentages

In order to enable direct comparison between stakeholder groups, and to avoid the confusion of the different sample sizes, it was decided to convert each group response, to a question, into a percentage of the group sample size. Unless otherwise stated, the percentages shown are the percentage of the total number of respondents in each group.

Where respondents were asked to rank items in order they are scored as follows:

Question B.16.

Within each stakeholder group.

The percentage of the total score The group score for this source X 100 for each information source = The total group score for all sources

Question B.18.

Within each stakeholder group.

First choice scores, 3, Second Choice, 2 and Third Choice, 1. The percentage of the total The group score for this measure X 100 score for each measure = The total group score for all measures.

Question B. 19.

Within each stakeholder group.

First choice scores, 5, Second, 4, Third, 3, Fourth 2 and Fifth 1. The percentage of the total score The total group score for the proposal X 100 for for each proposal. = The total group score for all proposals

DISCUSSION OF RESULTS

In this section the results, are summarized, and then their significance discussed.

The Experiences of the Public Question A1. What were the Main problems you experienced?

The responses were wide ranging, but can be summarized as follows. The EDM processes experienced were felt to be confrontational, lacked balance, and lacked real consultation. It was felt that the short time scales for public participation did not allow for effective and constructive engagement with the process, or the people working within the regulatory bodies. Some felt that there seemed to be another separate process at work, from which the public was excluded.

Question A2. How did you feel about your ability to influence the decision-making process, at the beginning and at its end?

The general trend here, with some exceptions was to start out sceptical but hopeful, and end up cynical and angry.

Question A3: Would you see the outcome of your campaign as positive?

Whilst only 40% responded Yes, all but one felt it was worthwhile becoming involved.

Question A4: What changes in the EDM process would have made you more content with the outcome?

The desire for proper consultation, within a reasonable time frame, based on good science, and backed by legal sanction, was widespread. There were many requests for impartiality of EIS, greater transparency, and for reasons to be given for the decisions made. Better-designed documents coupled with easy access to assistance in understanding same, were also requested.

Question A5. Are there any lessons you learned which would be useful to hand on to others starting a campaign?

Be positive, vigilant, steadfast, inclusive within your group, and well organised. Get good advice and educate yourself. Broaden the agenda, and be 'for something' rather than against. Establish good relationships with the officials you are dealing with. Do all of the above, but don't rely on the participatory process. The media/political campaign is also important.

Question A6. Did you understand the EDM process; Yes

a) At the beginning. 35%

b) At the end. 65%

Question A7. Did you receive assistance in understanding the process.

Yes 27%

Question A8. Were the time constraints for responses to proposals sufficient?

No 70%

Question A9. In your experience was it easy to access documents from; Yes

a) Local authority. 50%

b) EPA. 43%

c) Department of the Environment and local Government. 57% d) An Bord Pleanala. 60%

These are the percentages of those who actually contacted the relevant bodies.

Question A10. Did you believe the EDM process was legitimate and consistent with the aim of environmental protection?

Yes No No Reply 17% 53% 30%

Question A11. Did the EDM process you experienced increase your trust in the decisionmaking bodies? (Please respond only to those relevant to your situation).

a) Local Authority-Planning.

b) Local Authority-Waste Permitting.

c) E.P.A.

d) An Bord Pleanala.

e) Others (please name).

Question A12. Were you given an explanation as to the nature of the final decision, and the reasons for choosing the particular course of action?

Yes 37% No 43% No Reply 20%

Question A13. Did you feel your input was listened to?

Yes 30% No 60% No Reply 10%

Question A14. Did you feel that the responsible body had the competence to enforce the decisions?

Yes 38% No 38% No Reply 24%

Question A15.

Question A16. Please estimate the total cost to you of your involvement in the EDM process.

Answer: Costs in one year ranged from □70 to □10,000. Yes

(a) Did you find it easy to locate legal expertise? 10%

(b) Did you find it easy to find scientific expertise? 35%

(c) Did you find it easy to finance this expertise? 5%

(d) Was fund raising a major part of your campaign? 45%

Question A17. Did you spend more time than you expected on the process? Please give a rough estimate of the number of hours.

Yes 70%

Question A18. If there was an EIS;

Yes

a) Did it include all the main environmental impacts? 0%

b) Did it look at all the main alternatives? 0%

c) Did it give baseline data against which future impacts could be measured? 10%

d) Post project, are the EIS predictions being monitored and are the results publicly reported? 0%

Discussion of Section A. The experiences of the Public Half of the 20 Public respondents became involved in EDM processes as a result of particular issues that arose in their communities, and they then attempted to engage with the relevant regulatory authority. The other half became involved with longer-term issues, also in their communities, but which required on-going involvement over long periods, up to five years in one case.

Whilst 40% saw the outcome of their involvement as positive, and all but one saw it as a worthwhile learning exercise, the overall experience of dealing with proponents and regulators has generally been negative. Despite starting from a broadly constructive point of view, these respondents ended up largely cynical, frustrated and angry.

Only 17% of respondents believed the process in which they were involved to be legitimate and consistent with the aim of environmental protection. Their trust in the various regulatory agencies, following their experiences, was also well below acceptable levels, with zero for the EPA. From Question A.11, it can be seen that 75% of the respondents had experience of the planning process, but only 35-40% had experience of the other processes. Some of the reasons for these opinions can be seen from the responses to other questions.

There was a broad desire expressed for proper consultation. These desires were met with tight time constraints and minimal help in understanding the process involved. In some cases the treatment received was a hindrance in trying to get to grips with procedures. Only thirty percent felt they were being listened to, and only 37% had the reasons for the final decision explained to them. According to the respondents, none of the EIS studied by the respondents involved fulfilled the basic guidelines of the EPA, and no public post-project monitoring was carried out.

38% felt that the responsible body was competent to enforce the decision made. Only 43-60 % of respondents, depending on information source, felt access to documents was easy, despite the relevant legislation.

Add to this that, at the out set, only 35% understood the process they were engaging in, let alone the intricacies of the issues involved; that very few found it easy to locate legal or technical expertise, and that paying for these meant that, for 45%, fund-raising became a major part of their campaign. Individual expenditures were huge, ranging from □70 to □20,000 in one year, extending in some cases over many years, coupled with time spent of 70 hours total, to 15-20 hours per week again over years.

So what is the overall scenario emerging?

Here we have twenty individuals, most of whom are linked to groups with shared values, who attempt, as they see it, to protect their local environment, whilst perhaps trying to affect the bigger picture.

Apart from one laudable exception, it seems that they were not greeted with open arms. What should have been a positive first entry for many into the processes of EDM has instead left them cynical and angry, with one respondent opting out for the future, calling instead for nuisancemaking, street politics and media events, in order to create political pressure.

Here was a respondent who, in his words, started with optimism and campaigned constructively, attempting to engage with the regulators and proponents alike, and who found this engagement futile. The rest of the respondents were not far from sharing this opinion.

Finally, from Question 18, it is clear that the quality of EISs, that accompanied the EDM processes experienced by the respondents, was generally perceived to be of a low standard.

Attitudes of the Main Stakeholders

Question B1. Should the public be excluded from any particular area of EDM?

(Question for all stakeholders)

5% of the public, 14% of industry, 23% of EAOs and 28% of council respondents answered Yes, and of these all but two qualified their responses. In general the qualifications ruled out involvement in areas from which the public are already legally precluded.

There were some who saw the public as ill-informed and their involvement in EDM as inappropriate. Over all, though, there was only two who gave an unqualified Yes to this question, so it can be concluded that most of the respondents think that the public should not be included.

Question B2. Do you think that public involvement in EDM;

(Question for all stakeholders)

EAOs Industry Public Council

a) Leads to greater transparency? 95 73 80 100

b) Increases environmental protection? 95 75 78 50

c) Reduces conflict? 70 53 75 75

d) Increases public trust? 85 60 75 83

e) Slows down the EDM process? 55 73 55 92

d) Improves project design? 75 47 68 58

Figures given are % responding Yes.

Question B3. How do your staff react to Public Involvement in EDM?

Are they;

Industry EAO Council

a) Enthusiastic? 73 63 67

b) Cynical? 27 56 17

c) Apprehensive? 53 63 50

Figures given are % responding Yes.

(Asked of all stakeholders, but the public).

Discussion of Question B1-3

Attitudes of Main Stakeholders

Asked if the public should be excluded from the particular EDM processes, the strong concensus was against exclusion. Only 4% gave an unqualified yes to this proposition. There was a majority opinion that public participation in EDM leads to greater transparency, increases public trust, reduces conflict and improves project designs. The Industry respondents were the least convinced of the reduction in conflict and the improvements in design that might result.

There was also broad agreement that public participation would cause a slowing down of the EDM process. It was also clear that the staff of the Industry and Local Authority respondents were generally more enthusiastic than they were apprehensive, and that cynics were in the minority, except in the view of the EAOs, amongst the colleagues.

Industry Opinions

Question B4. Should the public have a greater involvement in environmental impact assessment (EIA) process?

Yes 57%

No 37%

No reply 6%

Question B5. If your answer was yes to the above at what stage(s):

a) Scoping?

b) Baselines?

c) Impact evaluation and prediction?

d) Mitigation planning?

e) Comparison of alternatives

f) Decisions on alternatives

g) Preparation of EIS. (industry stakeholders only)

Of those answering Yes to greater involvement, the following percentages gave positive reactions:

a) Scoping 75%

b) Baselines 33%

c) Impact evaluation and prediction 67%

d) Mitigation planning 43%

e) Comparison of alternatives 89%

f) Decisions on alternatives 71%

g) Preparation of EIS 33%

Discussion of Question B4-5 Industry Opinion

This section deals specifically with industry attitudes to public involvement in the EIA process.

61% of the respondents replying to this question said yes, the public should have greater involvement. This should be compared with the 95% who thought that public involvement would increase environmental protection and lead to greater transparency, as well as the 75% who saw it improving project design. (Question B.2). Of that 61%, a substantial majority were in favour of increased public involvement, in scoping, impact evaluation and prediction, comparison of alternatives, and the decision on which alternative to choose. The more technical areas of baselines, mitigation planning and preparation of the EIS all received much lower percentages.

Comparison of alternatives was the only category receiving majority support of the total response to B.4, i.e. 56%.

In essence then, a majority of the Industry respondents are in favour of increased public involvement in the EIA process but there is less clarity about what it is they should be involved in. The stage of 'comparison of alternatives' being the only one with an overall majority. From this the conclusion may be drawn that public involvement is cautiously welcomed but only in the generally non-technical area of comparing alternative proposals, without necessarily being involved in the decision-making that follows from this. Given the positive general responses that Questions B.2 and B.3 received, it would seem that (in the context of the EIA process) the industry respondents like the idea of public participation, but not necessarily the reality.

Experiences of the Main Stakeholders

Questions addressed to all stakeholders but the public.

Question B6. Public involvement in EDM has;

Industry EAO Council

Been a positive experience. 67 88 83 Used a lot of staff time. 47 44 75 Been very expensive. 20 25 9 % answering Yes.

Discussion of Question B6.

All three groups gave strong affirmative responses to having had positive experiences of public involvement in EDM. Only the Council officials gave a strong response to a lot of staff time being used, although it was a significant factor for all three.

The financial costs of public involvement did not seem to be a major concern for most respondents. The numbers who gave no response to each question are also interesting. To each question in sequence there were 20%, 26% and 27% for industry. One explanation could be that they had no experience of public involvement. With regard to use of staff time and the costs involved it would appear, perhaps not surprisingly that fewer EAO's responded than Council officials, perhaps because the latter are more likely to know the answers to these questions.

Capacity For Public Participation

Questions directed at all stakeholders except the public.

Question B7. Are procedures for dealing with requests from the public, for information on the environment and EDM, strictly laid down?

Yes(%)

Industry. 40%

Local Authority. 58%

Question B8. Do you provide assistance to the public in understanding;

a) Legislation and its implementation?

Councils EAO

Yes 83% 88%

b) Technical documents and their relevance?

Councils EAO Industry

Yes 83% 69% 33%

Question B9. What pro active methods do you use to disseminate information to the public on;

(a) general environmental issues?

(b) specific issues?

Questions B9 and B10 are answered together, below, as the responses overlapped regularly.

Question B10. What pro-active methods do you use to promote feed-back from the public on environmental issues?

Answers to B9 and B10 Industry.

In general the response showed a passive approach to information giving, though a minority looked for feedback from their surrounding communities.

EAO.

As would be expected, it being their job, the EAOs were the most pro-active in this area, and use a wide range of feedback mechanisms. Council.

More traditional use of the media, public meetings and site-visits were coupled with a heavy emphasis on the use and development of web-sites. In general this group favoured a passive approach, but with a willingness to engage with the public.

The Department of Environment and Local Government also emphasized its website development, which it saw in part as a screening agent, i.e. the website enables the public to understand the workings of the department, so enabling them to ask clearer and more focused questions of department staff, and consequently facilitating them in their quest for the exact information they need, whilst reducing time spent by staff trying to understand the questions asked.

Question B11. Does your Local Authority website have links to any of the following;

Local bye-laws? 50%

National legislation? 46%

European directives? 29%

ENFO? 46%

Question B12. Does your department have the capacity to give good access to information on the environment?

Yes(%) No(%) No Reply(%)

Council 92 8 0

EAO 84 9 7

Question B13. Which of the following files are now available on your website;

Planning applications? 54%

Waste permits? 45%

Discharge to water licences? 28%

Water monitoring reports? 35%

Air monitoring reports? 25%

Noise monitoring reports? 8%

Litter hot spots? 8%

Prosecutions (ref environmental legislation)? 25%

Percentage Responding Yes.

Question B14. Do you have a guaranteed call-back information hot-line?

Industry Local Authority

Yes 40% 25%

Question B15. Are your staff trained in the following public access skills?

IndustryEAO Council

Public meetings 47 88 67

Workshops 47 81 33

Alternative Dispute Resolution 27 31 25

Media Presentations 53 69 50

Advertising 34 50 33

Writing articles/newsletters. 47 75 58

Percentages responding Yes.

Question B16. What outside sources of information do you use?

Please rate each, according to your experience, from 0-5 with 5 being excellent.

a) ENFO.

b) Local Authority.

c) EPA.

d) Library.

e) "Expert".

f) Third Level Institute.

g) Other.

h) DOELG

Discussion of Question B6-16

Capacity for Public Participation

From the responses to B6 and B7 it is clear that 60% of the industry do not have formal lines of communication with the wider community or help with comprehension of technical documents, and are generally passive in this regard.

By contrast, Local Authority (LA) staff do see a strong role for themselves in providing assistance, and in understanding legislation and technical documents, even though only 58% have clearly laid down procedures for requests for information. Even then the procedures are generally only in relation to requests under the Freedom of Information Act 1997. Only 27% of LA respondents have a freephone call-line, compared to 40% in industry. LA staff assess their own capacity, to give good access to information on the environment, very highly. (B11) Looking at B8 and B9 it is clear that the council staff are inclined towards the more traditional, generally passive methods of information giving but with a strong reliance on website development. EAOs by contrast and by the nature of their job, are more pro-active and use a wider range of tools. Many local authorities are still developing their websites, and so the availability of information on the different LA websites varies dramatically in quality and quantity, as can be seen from responses to B10 and B12.

B14 responses showed a generally low public access skills base, with not surprisingly, EAOs being better trained overall.

The sources of information were used by all the stakeholders was the subject of B16. The Public generally used independent sources, although surprisingly few used the excellent resources of ENFO. The use of the local authority, as a source, received a rating of 20% of the public total. Council staff also made little use of ENFO, sourcing most of their information with the EPA and the DOELG with some use of consultants. By contrast then ENFO was the first choice of EAOs, with the EPA and DOELG second.

Reflected here may be the lack of trust felt by the public in the regulatory authorities. It should be noted that the DOELG was not offered to the public stakeholders as a choice. The Local Authority is a necessary source of some information for most EDM processes and so received a larger number of individual public responses than the other sources, but its average score was 2.4 compared with 4.8 for "other" and 3.8 for "experts".

To summarise, the capacity to provide good access to public participation is at best patchy, despite the response by local authorities, to B11, asserting c.90% capacity to give good access to information. Whilst the inter-net is an invaluable public access tool for public bodies and industries, the existence of the 'digital divide' is not mentioned by anyone. Most of the websites are still at an early stage of development, and very few are interactive. Yet the inter-net seems to be the main focus for the council respondents in particular.

Capacity Building

Question B17. What changes in your company/department would improve public access to;

a) information on the environment?

b) Environmental decision-making

Answer and discussion.

The response of the Industry respondents seemed to be generally aimed at increased transparency and was marked by a desire to reach out to the community. Both of the Local Authority respondent groups were looking for increased resources, staff training and improved websites. Increased public access to Strategic Policy Committees and regular meetings with key stakeholders were both good capacity building suggestions. A minority in both EAO and Council respondents felt no change was necessary to improve access to decision-making, with two EAOs suggesting that access to county councilors was sufficient. Question B18. What capacity building measures for public involvement in EDM would you like to see developed in your organization? (Please select and rank your top three)

a) staff training on legislation and technical details.

b) increased ease of access to information for staff.

c) staff training in public access skills.

d) increased advertising budgets.

e) appointment of a public information officer.

f) training in web-site development.

g) other/s (specify).

Question B18 looks at in-house capacity building in Industry and Local Authority.. All the stakeholder groups seem to focus on skills training and access to information for their own staff. Industry looks for staff training in legislation and technical details, as do EAOs, whereas the main priority for council staff is training in participation skills and greater ease of access to information for staff members.

EAOs in particular want increased budgets for advertising, which is a key part of their work.

Question B19. The following is a list of proposals designed to improve public participation in EDM..

Please select the five most useful, and rank them with no.1 being your first choice etc. Percentages shown represent the cumulative percentage scores from each of the stakeholder groups.

Cumulative

% Scores.

A community based, independent environmental information broker. 37%

An environmental ombudsman. 58%

A one-stop information hotline with guaranteed call back. 29%

Technical assistance grants. 28%

Science shops*. 13%

Citizen training in how to use the EDM processes. 33%

Collaborative participation where all stakeholders (including developers and concerned citizens) share information and decision-making. 60%

Increased availability of information in various formats. 49%

Grants to community environment groups for educational purposes. 32%

Improved access to documents. 26%

An independent 'expert' panel for review of technical information. 35%

Total: 400%

*Science shops do community-based research where citizens set the agenda. University or college based, the talents of the students, supported by faculty advisers are used to answer questions posed.

This question is based on a list of measures, designed to build the capacity of the public for participation in the activities of the US EPA. (Breggin, L. and Hallman, H.)

Discussion of Question B19.

The results for this question are based on the cumulative % scores for each stakeholder group.

This was done to avoid distortion of the bar diagram due to the different sample sizes in each stakeholder group. Caution should be used in interpreting these results. (The percentages given here are of the score within each stakeholders group). Collaborative participation' was the overall first choice followed by the 'ombudsman' and then 'increased information'.

The Publics first choice was 23% for collaborative participation, with the environmental broker and the ombudsman being equal second choice at 16%.

Industry gave 24% of its overall score to the environmental ombudsman, 14% to increased information and 10% to the broker.

Council respondents gave their highest score (14%) to a "one-stop hot-line" with collaboration

13% and increased information 12%. They also gave 11% to each of 'technical grant assistance', 'improved document access' and 'expert panel'.

EAOs scored 17% for 'increased information availability', 15% for ' collaboration' and 11% each for 'environmental broker' and 'ombudsman'.

Combining the percentage scores for the 'broker', 'ombudsman', and 'expert panel', captures 44% of the Industry score, 41% of the Public score, 27% of the EAO score, and only 18% of the Council score.

In summary then, from the above, it is clear that there is a consensus among the stakeholders for working together in EDM processes. At the same time, there is a recognition that some issues need arbitration, as well as independent assessment of disputed 'facts'.

As it is usually proponents of projects and the public that are in dispute, it is not surprising to find them opting for arbitration whereas the council officials are looking more at the practicalities of file access, hot-lines and the availability of information, including technical information. The EAOs, not surprisingly since it is part of their job specification, scored highly for increased availability of information.

Question B20. Are there any lessons you have learned, about involving the public in EDM which might be of use to other public bodies?

Answers and discussion.

Industry. It was felt by two of the respondents that the social and political climate was not mature enough for public participation (PP). Others called for openness with a focus on working with small groups.

Councils. Generally, suggestions given were of a positive nature. It was suggested that the public should be brought in as early as possible to EDM processes. It was also suggested that PP be made an integral part of all EDM processes. Even if processes are slowed down, you cannot, according to one respondent, consult too much.

EAOs. PP in EDM requires openness and patience, and is best done face to face. The public should be made aware of the constraints on the council. People are only interested in contentious issues. Involve other community bodies.

RBM. Have sufficient resources at the outset to conduct your participatory process. Make sure, that the public are aware of the limits of the process, at the beginning. Make use of trained facilitators.

DOELG. Make people aware of their right to access information on the environment. Discussion.

Most of the responses were positive towards PP, though the DOELG talked only of access to information. Those that pointed to the political immaturity, and the interest in only contentious issues, pose the question in the author's head "how is this situation going to change?"

Summary and Discussion of Results

Given the inevitable drive to build a sustainable society in which the economic sociologic and environmental aspects of development are all considered, the importance of including all the stakeholders in environmental decision-making (EDM) becomes increasingly apparent. All actions for change have consequences, and the moves to greater public participation have their costs as well as benefits. It is necessary for all people to decide whether these costs are greater or lesser than the costs of no change. In order to make that choice it is essential to have as much information as possible. This study attempted to draw together some of the arguments for and against that change. Making the assumption that some changes are coming already, as a result of the Aarhus Convention and its offspring, this study has examined ways of making public participation in EDM effective.

In understanding any change, it is necessary to know where you are starting from, and the possible routes that might be traveled to get there. To this end a survey was made of the major stakeholders in EDM processes, industry, the regulators, and the public, in order to see the starting point, as well as some of the possible ways forward.

In the time-scale of this treatise, it was not possible to get statistically representative samples of either industry or public opinion, although the local authority sample was sizeable. 35% of environment departments, 47% of EAO's covered in total 65% of Local Authorities. The methods by which the public and industry groups were contacted were very specific, so this and their self-selection in responding should be taken into account in interpreting the results. Every ENGO on the ENFO list was surveyed, but only one member of one organization responded, despite each receiving two reminders. Several said they did not have the time.

The authors impression was that they mostly saw public participation as an irrelevance to them. The IFA (Irish Farmer's Association), I.B.E.C. (Irish Business and Employers Confederation) and all the government departments with an obvious environmental remit were also surveyed to no avail, apart from the DOELG, who did respond.

From this the conclusion could be drawn that either these bodies are not interested in the survey, or they are not interested in public participation, or they see participatory practices as a dilution of their power. This is an area for future research.

In order to get a reasonable return on the survey, it was necessary to send out the questionnaire long before the literature survey was completed. This meant that some questions that should have been asked were not, and others were asked badly. An example of the latter would be Question B.1.-Should the public be excluded from any particular area of EDM? The strong response against exclusion does not, however, indicate whether they think the public should be included. With regard to the former, it would have been helpful to find out why each group thought public participation should entail.

Taking the above reservations into account, and in the context of the literature survey, a number of indicators for future action can be taken from the results of the questionnaire survey, as will be seen. The questionnaire looked in turn at the experiences and opinions of the stakeholders, as well as their capacity to deliver public access to information and EDM. Opinions on capacity building measures were also sought.

Putting together the experiences of the different stakeholders, it fast becomes apparent that they have been very different.

On the one hand, we had the public stakeholders, who clearly felt that they have been largely ignored despite their best attempts at constructive engagement with the regulatory authorities. Their commitment cannot be denied in both time and money, and yet they seem to have met, in general, with unhelpful and sometimes obstructive treatment by public servants at local and national level. On the other hand we heard from those people working in Local Authorities and the DOELG that they were confident in their ability to provide good information to the public, and that although it might slow EDM processes down, they had the majority opinion that public participation had many positive attributes. They also felt that their staff were enthusiastic (c65%) though apprehensive about public involvement in EDM. Industry respondents were even more enthusiastic for public involvement, though they did not have the same strength of belief in its ability to reduce conflict or improve project design.

Clearly something is wrong here. Assuming that all the respondents are genuine in their responses, and that it would appear that they would all wish to operate in a more open way, with the protection of the environment as an end point, there is clearly either a communications breakdown, or else the structures and laws set up to protect the environment are unable to deliver public participation.

It seems like a mixture of the two. In order to communicate well, and under time pressure, it is necessary to speak the same language and to know the rules of engagement. It seems that one third of the public did not understand the EDM processes they were entering into, and only two thirds understood at the end, with very few (27%) getting any assistance along the way. And yet the public servants seem more than willing some (c80%). They were also willing to help with technical matters (c75%).

Independent expertise was hard got by the public and expensive, and yet in situations where the proponent of the project or plan was a government body, they saw themselves as paying for expertise on both sides of the debate.

What happens when a member of the public requests information? Only 58% of Local Authorities said they had strictly laid down procedures and only 40% of industry. Where guaranteed call back hotlines operate (only 25% of Local Authorities) response times were from twenty minutes to 72 hours.

It would be useful here to know how the front-of-house people in Local Authorities feel about their ability to deal with these questions. Do they have the training, back-up and staff numbers enable them to give the service they would like. These questions, unfortunately, were not put. However, in looking at questions relating to capacity, the following becomes clear. For enabling information delivery, the web-site is a powerful tool but limited in its accessibility, and yet it seems to be the focus of most of the council respondents. At the same time, despite some shining stars, environmental information on Local Authority web-sites, is still fairly restricted, as in that on the

E.P.A. and DOELG web-sites. However; they are all in the process of improving, and in the case of the latter two, undergoing radical overhauls. Links from Local Authority web110 sites to other sources of information ranged from 50% for local bye-laws, down to 29% for European Directives.

The author is not aware of any research into who accesses the existing sites, or who knows that they exist, and yet they seem to be the main focus for government resources, and local government officials, in terms of improving accessibility.

Whilst EAO's, whose job is to educate and interact with the public on a daily basis, rate their public access skills highly, they are not the ones on the front line in EDM processes. The percentages of Industry respondents in particular and Council respondents that replied positively to having public access skills was low in general, except in media skills, and for the council officials with regard to public meetings.

It is not to be expected that these two groups would be highly trained in these areas, it is not their first priority, but these skills, particularly the facilitators skills needed to work with small groups and in workshops, should be available, in house, in every government agency or Local Authority dealing with the public in EDM.

These skills would also be useful in other aspects of their internal and external business, e.g. Partnership and the development of E.M.A.S. Generally though, when asked about proactive methods of giving information and receiving feedback, industry was very passive in its approach. Council officials also tended to be passive but showed a greater willingness to engage with the public. As mentioned earlier, the web-site seemed to be the tool of choice for council officials.

The EAO's and the DOELG were both very proactive in information giving and feedback but the EAO's as educators showed the widest range of skills.

Access to outside sources of information is another area of capacity that needs examination. Question B.16. showed council officials, who deal largely with the technical/decision making aspects of EDM, opting for sources of information that enable them in these tasks. Similarly EAO's, who used the educational resources of ENFO also used the DOELG and EPA. By contrast, the public largely used independent sources which got very high average scores of 4.8 111 for 'other' and 3.8 for 'experts' compared with only 2.4 (out of 5) for the Local Authority, even though 75% of the respondents were involved in EDM processes to do with Local Authorities and planning. None of the public respondents were from Dublin, which might explain the low use of ENFO (8%). Those that did, gave it a score of 3.2 on average. The author would score ENFO much higher as a resource, from his own experience, and would recommend that ENFO be resourced such that it can make its presence felt in every region of the country.

The use of independent sources of information by the public would tie in with Question A.11 where only one third of the 75% of respondents dealing with local government planning issues had increased trust at the processes end, and a lower figure dealing with waste permitting (22%). Trust in the EPA was zero and in An Bord Pleanala 43%.

The only area of actual EDM discussed in the questionnaire was the EIA process. Clearly the public respondents that had had experience of EIS's found them of very poor quality indeed. Alongside this, we have the Industry respondents, who were 73% enthusiastic about public involvement in EDM, giving only 57% support for greater public involvement in the EIA process, with 53% agreeing with public involvement in the 'comparison of alternatives'.

So this is the base line, we had a largely disgruntled public, an enthusiastic but under-skilled and under-resourced public service, and an industry response that was generally lukewarm. How did these same respondents see the way forward, what capacity building did they see as useful? When asked what changes in their organization would improve public access to environmental information and EDM, Industry respondents look to increase the transparency of their enterprises, and showed a desire to reach out into the community.

Local Authority and EAO respondents both looked for increased resources, staff training and improved web-sites. Regular meetings with key stakeholders in the community, coupled with increased public access to Strategic Policy Committees were both creative suggestions.

When asked to make choices about specific capacity building measures, Council respondents opted for staff training in public access skills and increased access to information for their staff; whilst the main choices for industry were staff training in legislation and technical details alongside training in public access skills. EAO's looked for advertising budgets to assist with their education programmes.

The penultimate question, B.19, dealt with capacity building measures for the wider community,.

The measures receiving the highest cumulative scores were; 'collaborative participation where all stakeholders share information and decision-making' at 60%, and 'an environmental ombudsman' at 58%.

Apart from this, Industry and the Public both scored very highly for the three interventionist measures of; an expert panel to resolve disputes over 'facts; a community based independent environmental information broker, to give that expertise sought by the public; and an ombudsman to act as a final arbiter if all else fails.

The final question B.20, asked of all, but the public, looked for advice to others, based on the experiences of the respondents. Most of the responses were positive to PP, suggesting a more personal approach to the EDM processes with PP incorporated into the EDM process at the earliest possible stage. Two responses from Industry felt the politico/sociological climate in Ireland was not mature enough. Perhaps by reaching out into the communities, of which they are but one part, they might be surprised. There is a great wealth of detail in the responses of all the stakeholders, which can be seen in Appendix II and which in its variety emphasizes the fact that, as in all human activity, nothing is black or white.

CONCLUSIONS

It is clear that the present situation with legislation, that largely restricts public participation to 'report and comment' activities, is likely to leave the public stakeholders with feelings of frustration and alienation.

It is also clear, that without the necessary changes in practice, and increases in resources for skills training and more personnel, public servants cannot deliver effective access to the public.

The lack of trust shown by the public in the regulatory authorities, put together with the apprehensions of the regulators and industry, about public involvement in EDM, will require a great deal of work to change. Open and inclusive debate, amongst all stakeholders, about what changes are needed in their inter-relationships and in the distribution of decision-making power are essential.

The Directives coming from Europe and emanating from the Aarhus Convention should be seen as only the minimum standards, and the spirit of Principle 10 brought in when they are enacted into Irish law. Indeed, the drafting of new legislation should itself be open to public consultation.

Greater public awareness of their rights to be involved in EDM processes, and greater knowledge of how to be involved would, it was agreed, lead to the end-game of greater environmental protection. The more the public are involved in EDM, the better educated they will become on environmental issues, and hopefully then the safer the environment will be from thoughtless plundering and degradation of natural resources.

The way forward then lies in; greater openness, a willingness, on the part of regulators and industry, to reach out pro-actively into their communities, a detailed assessment of the capacity needed to deliver participation, and a commitment on behalf of all stakeholders to work together for their mutual benefit, and a sustainable future for all life in this Biosphere of which we are a small but potentially very destructive part.

RECOMMENDATIONS

- That Ireland, as a signatory, actively seeks to carry the spirit of Principle 10 into all areas of EDM.
- That Irish Governments, in drafting legislation on the environment, proactively consult all stakeholders beforehand.
- That the Irish Government finance an assessment of Principle 10 implementation, as agreed under Partnership for Principle 10, using an independent assessment method such as the Access Initiative. In this way, capacity building will have a base-line to focus on.
- That regulators proactively seek out stakeholders and encourage their involvement in EDM well before the final decision is made e.g. calling door to door in the neighbourhood of a proposed development.
- That timescales for submissions about EDM should be more flexible, allowing more time for the consideration of complex issues.
- That funding be made available, to enable public participation, in the form of grants for legal and technical assistance.
- That the public be involved as early as possible in the EIA process. Preferably from the screening phase onwards, sharing in the information flow that leads to the EIS.
- That reasons be publicly given for the final decisions in all EDM including the decisions of An Bord Pleanala, and that responses to submissions are included with same.
- That the role of ENFO should be expanded both geographically and in terms of its service provision. This latter to include an independent environmental information broker, with access to a panel of independent experts.

BIBLIOGRAPHY

Acquaah, George: *Understanding Biotechnology: An Integrated and Cyber-Based Approach*, NJ, Pearson/Prentice Hall, 2004.

Ausubel, F.M.: *Current Protocols in Molecular Biology*, New York, John Wiley and Sons, 1989.

Baker, Kenneth Frank: *Ecology of Soil-bourne Plant Pathogens, Prelude to Biological Control*, Berkeley, University of California Press, 1965.

Bauer MW: *Biotechnology-the Making of a Global Controversy*, Cambridge, Cambridge University Press, 2002.

Bernauer, Thomas: *Genes, Trade, and Regulation: The Seeds of Conflict in Food Biotechnology*, Princeton, Princeton University. 2003.

Broach, J.R.: *The Molecular Biology of the Yeast*, Cold Spring Harbor, Cold Spring Harbor Laboratory, 1981.

Chrispeels, Maarten : *Plants, Genes and Crop Biotechnology*, Sudbury MA, Jones and Barlett Publishers, 2003.

Costanza, R.: *Ecological Economics*, Columbia Uni. Press., New York, 1991.

Daubenmire, F.: *Plants and Environment*, New York, Wiley, 1947.

Dusenbery, David B.: *Life at a Small Scale: the Behavior of Microbes*, New York, Scientific American Library, 1996.

E. Dereniak and D. Crowe, *Optical Radiation Detectord*, John Wiley and Sons, 1984.

F. Reif: *Fundamentals of Statistical Mechanics*, McGraw-Hill, Singapore, 1965.

Fransman M, Junne G, Roobeek A: *The Biotechnology Revolution?*, Oxford, Blackwell, 1995.

Fumento, Michael: *Bioevolution: How Biotechnology is Changing Our World*, San Francisco, Encounter Books, 2003.

Goodman, D.C.: *From Farming to Biotechnology: A Theory of Agro-industrial Development*, Oxford, Blackwell, 1987.

Jameson, J. L.: *Principles of Molecular Medicine*, Totowa, Humana Press, 1998.

Li, Koibong and Leslie A. Lewis.: *A Novel Approach to the Evolution of Microorganisms*, New York, Vantage Press, 1987.

Lodish, H.: *Molecular Cell Biology*, New York, Freeman, 2000.

Lynch, J.M., and Hobbie, J.E.: *Micro-organisms in Action: Concepts and Applications in Microbial Ecology*, Blackwell Scientific Publications, Oxford, 1988.

Maarten, C.: *Plants, Genes and Crop Biotechnology*, Sudbury MA, Jones and Barlett Publishers, 2003.

Metting, Jr., F.B.: *Soil Microbial Ecology: Applications in Agricultural and Environmental Management*, Marcel Dekker, New York, 1992.

Raff, K. Roberts, and J. Watson, *Molecular Biology of the Cell*, New York: Garland Publishing, 1989.

Rietman, Ed.: *Molecular Engineering of Nanosystems*, New York, Springer, 2001.

Sayler, G.S., Fox, R., and Blackburn, J.W.: *Environmental Biotechnology for Waste Treatment*, Plenum Press, New York, 1991.

Sheldon, J.K.: *Practical Environmental Bioremediation*, Lewis Publishers, Boca Raton, 1992.

Switzer, R.L.: *Experimental Biochemistry*, New York, W.H. Freeman and Company, 1977.

Teich, M.: *Fundamentals of Photonics*, New York: Wiley, 1991.

Wood, E.J. Ferguson: *Marine Microbial Ecology*, London, Chapman and Hall, 1965.

Yanoff, M.; Duker, J. S.: *Ophthalmology*, St. Louis, Mosby, 2004.

INDEX

R

S

T

W

□□□